THE WHICH? GUIDE TO HI-FI

THE WHICH? GUIDE TO HI-FI

Published by Consumers' Association
and
Hodder & Stoughton

The Which? Guide To Hi-Fi
is published in Great Britain
by Consumers' Association
14 Buckingham Street, London WC2N 6DS
and by Hodder & Stoughton
47 Bedford Square, London WC1B 3DP

Written by	Quentin Deane
Edited by	Brian Guthrie
Design Consultants	Turner Wilks Dandridge Limited
Illustrated by	Adam Banks and Tom Cross
Cover by	Brian McIntyre (Ian Fleming Associates)

The publishers thank, for help in the preparation of this book,

BBC Engineering Information Department
Martin Colloms
Gordon King
Angus McKenzie MBE
Paul Messenger
RS Roberts
MG Scroggie
Sue Thomas

First edition, first reprint
© Consumers' Association 1982

All rights reserved.
No part of this book may be reproduced,
in any manner whatsoever, without prior
permission from the publishers.

ISBN 0 340 26629 5

Typeset by Vantage Photosetting Company Limited
North Baddesley, Hampshire
Printed and bound in Great Britain by
Redwood Burn Limited, Trowbridge, Wiltshire

Contents

Introduction		7
Chapter 1	Buying hi-fi	9
Chapter 2	All about sound	18
Chapter 3	Hi-fi fundamentals	25
Chapter 4	Loudspeakers	34
Chapter 5	Amplifiers	46
Chapter 6	Record decks and cartridges	59
Chapter 7	Radio tuners and FM aerials	80
Chapter 8	Cassette decks and tapes	95
Chapter 9	Reel-to-reel tape decks	112
Chapter 10	Headphones	118
Chapter 11	Microphones	122
Index		127

Introduction

Hi-fi is as complicated a subject as you want to make it. We want to make it no more complicated than you need to make good buys, and to get the most out of using those buys.

So this book is a mixture of more or less technical information – presented as simply and clearly as we know how – and down-to-earth advice. Much of that advice is about which of manufacturers' technical claims you can ignore, or take with a substantial pinch of salt.

There is perhaps no other range of consumer goods whose marketing is attended by as much technical and quasi-technical data as hi-fi. Are we to assume from the specification-filled hi-fi ad pages that different consumers buy hi-fi than buy fridges or cars? The answer, of course, is no. Can the workings of hi-fi somehow be inherently less easy to explain than those of washing machines? The answer, again, is a (qualified) no.

What has happened is that hi-fi manufacturers have, by and large, chosen the technical route to persuasion; aided and abetted, it must be said, by many of the pundits writing in hi-fi magazines.

This book is dedicated to cutting the mumbo-jumbo down to size.

That's not to say that you can ignore technical data. You really do need a certain, minimal grounding in it in order to appreciate what is important and what is not. This book is organised to make that process as painless as possible.

It starts with three introductory chapters.

Chapter 1, Buying hi-fi, is a simple guide to what is available – record decks; cassette and reel-to-reel tape decks; tuners; amplifiers; speakers; headphones; combinations thereof – music centres, hi-fi racks and so on. With it, you can decide what types you want to buy, where to look for advice, how much to spend and where to spend it. You may well return to this chapter when you've read the rest of the book – or as much of it as you need to read.

Chapter 2, All about sound, explains just what constitutes the physical phenomenon we call sound. Hi-fi is nothing more than a way of reproducing sound; this chapter sets the scene for understanding what it is hi-fi tries to do.

Chapter 3, Hi-fi fundamentals, explains the terms used to describe and measure hi-fi performance – and the tests that are commonly applied to many different types of hi-fi equipment. So this chapter provides the grounding for the specific performance information in subsequent chapters.

The remaining eight chapters deal with each of the main separate units of hi-fi equipment – loudspeakers, amplifiers, record decks, tuners, cassette decks – with smaller chapters on reel-to-reel tape decks, headphones and microphones. (Information necessary to assess the amplifier sections of tuner/amplifiers is contained in the amplifier chapter.) Within each chapter, there is information on:
- what the units do and how they work
- the various features they have, and how useful each is likely to be to you
- the technical tests that are done, and what the results mean – to help you interpret manufacturers' specifications, magazine reviews and so on.

Above all, there is information on how you can choose and use each piece of equipment, with sensible, direct buying information, and no-waffle advice. The paragraphs that contain this important advice are printed in bold type, as this paragraph is, to help you spot them easily.

Chapter 1 Buying hi-fi

To start with: what *is* hi-fi? Is it the same as stereo? **Hi-fi** is short for *high-fidelity sound reproduction* – recreating a sound that is as close as possible to the original sound. Most of the sounds that we want to re-create in the home are musical ones, but we may want to listen to speech, or natural or man-made sound effects as well. For convenience, this book uses the word *music* to mean all types of sound.

Stereo is short for *stereophonic*; this means, not two-dimensional sound as many people think, but solid sound. It is a method of reproducing music with the width and depth of the original, so that the musicians appear to occupy the same relative positions that they did when the recording was being made. With mono (monaural), the sound all comes from a single spot, and the musicians all appear to be sitting on each others' laps.

The image created even by stereo can appear a bit flat, with the musicians appearing to be in a straight line, and with the loss of much of the feeling of space that you might get in a concert hall. To reproduce some of this depth and ambience, many *quadrophonic* systems have been proposed in recent years. These have died a marketing death. A reincarnation, under the name *ambisonics*, seems likely. But don't let that deter you from buying hi-fi now.

So the terms hi-fi and stereo are not interchangeable. Knowing the difference may well save a very basic disappointment when buying equipment. As hi-fi means good and accurate, hi-fi equipment must be stereo (at least). On the other hand, stereo is simply one of the techniques for producing sound – so a stereo system need not be of hi-fi quality, and you should not assume that buying stereo means you are buying hi-fi.

This book is about hi-fi and, therefore, always about stereo.

Separates

In the good old days, audio equipment was very simple. You bought a large box called a radiogram, plugged it into the mains and possibly into an aerial and away you went – you could play gramophone records or listen to the radio, and the sound came out of a cloth-covered grille at the front of the box.

Even then, this apparent simplicity (by today's standards) disguised a number of different bits of equipment. In the last 15 years or so, the tendency has been to sell the bits separately. Each piece of equipment comes in its own box; you select the pieces individually (and to some extent ensure that they partner each other well); when you get them home, you connect them all together and plug some of them separately into the mains.

There are three main groups of these so-called separates. (You generally need one of each before you actually hear anything.)

● The **programme source**, or signal source, is the piece of equipment that actually reproduces the music. A *record deck* (with its associated stylus and pick-up cartridge) reproduces the sound held by gramophone records (which this book tends to call *discs*). A *radio tuner* picks up broadcast radio programmes. A *tape deck* plays back (and records on) magnetic tapes – either *cassette* or, less commonly now, *reel-to-reel*.

● An **amplifier** turns the tiny electrical signals from the programme sources into large ones, capable of driving loudspeakers. It is also a useful place to add switches allowing you to select which programme source you want to listen to, which you want to record from, and so on. It is also usual to have knobs to modify the signal in various ways – though these won't necessarily make the final sound quality better.

CHAPTER 1

● The **loudspeaker** is the end of the hi-fi chain, the unit that actually produces a sound you can listen to. It is fed the stepped-up signals from the amplifier (it won't work if connected straight to programme sources, because their signals are too small). For stereo, you need two loudspeakers (they are usually sold in pairs, anyway). The only alternative to loudspeakers is a pair of *headphones* – sometimes useful for private listening and, as a bonus, can usually be connected straight to a tuner or cassette deck (but not a record deck), possibly saving you the cost of an amplifier.

The rewards for suffering the complications of picking, choosing and installing separates can be high. You can build up gradually: perhaps a record deck first (plus amplifiers and speakers), then a tuner, finally a cassette deck. You don't have to buy equipment you don't intend to make use of – for example, if your main interest is in radio, you can spend all your money on a tuner (plus amplifier and speakers) and not bother getting a record deck or cassette deck at all. You don't have to stick to one manufacturer's products – and few manufacturers give the best value for money over the entire range of equipment. If one of the programme sources is faulty you can send just that piece to be repaired and still enjoy listening to the other bits. Upgrading (selling what you have got and buying better-quality equipment) is easier – you can do so piece by piece.

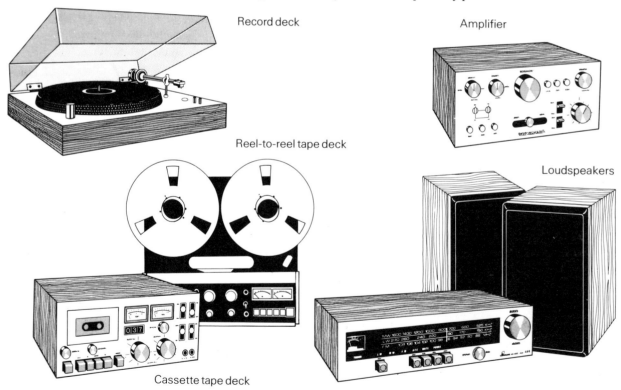

Record deck

Amplifier

Reel-to-reel tape deck

Loudspeakers

Cassette tape deck

Radio tuner

Combinations

But if you want to save yourself some of the problems of buying the equipment in the first place, then there are various combinations you can go for.

Music centres. Most have an amplifier, tuner, cassette deck and record deck all in one (rather large) box. The loudspeakers are separate – essential to allow you to place them in the room for best sound and good stereo effect. Paradoxically, the cheaper music centres often come complete with a pair of speakers: with the more expensive ones you may have to buy the speakers separately (or at least, you may have the option to).

Music centre

The cheapest music centres are incredibly cheap – the whole works need cost little more than the price of a single piece of even reasonable-quality separates equipment. But *Which?* tests have shown that the sound quality of such music centres is limited, to say the least. Adding better speakers would probably help (if you were willing to pay the price), but really these music centres are best used for providing background music or for use by very tolerant listeners. (Incidentally, do not make the mistake of assuming that young people are particularly tolerant of mediocre sound quality. In fact, their hearing is more acute than adults' and given the opportunity, they can be easily as critical as other people.)

There are, however, many real hi-fi music centres, with perfectly acceptable sound quality. These may cost little less than equivalent separates; so (providing you want all the programme sources that a music centre gives) your choice depends on whether you value the convenience of purchasing and installing a music centre more than the flexibility that separates provide.

If you can choose your own speakers, do so – they are likely to have the biggest effect on sound quality, so you do have the chance to influence quite strongly the sound that you hear. Do not think you are restricted to the same brand of speaker as music centre; indeed, few music centre manufacturers are particularly renowned for making speakers, so a good rule of thumb might be to start by assuming you need a *different* brand. Follow the advice in chapter 4 as fully as if you were buying loudspeakers for a separates system – including, if you can, the advice that you ought to choose your speakers before the rest of the system.

Hi-fi racks or towers are the latest hi-fi craze. They consist basically of a set of separates, all from the same manufacturer, with a tall cabinet to house them in and a pair of loudspeakers.

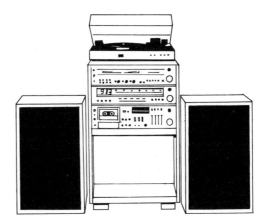

Hi-fi rack

There is usually a little more flexibility with rack systems than with music centres: there may be a (small) choice for each of the units; speakers are usually optional; you may not even have to buy all the types of unit. And, as the choices are fairly limited, you might think deciding on what rack system to buy is less taxing than deciding on a set of separates (though a little more so than with a music centre). Because all the units come from the same manufacturer, you may reckon they look nice together.

But racks have several drawbacks. You have to do as much installation work as with separates; upgrading is still difficult because the new units will have to be much the same size as the old ones; you are still stuck with buying everything (except perhaps loudspeakers) from the same manufacturer; and the cabinet may not be the best way to mount a record deck (chapter 6 says why this is important).

A rack system usually costs a little more (£30, perhaps) than the components bought individually – to cover the cost of the cabinet.

Tuner/amplifier. A tuner and amplifier in one box (a tuner/amplifier, or receiver) is probably the most common hi-fi combination. If you are starting from scratch and want a tuner in your set-up, going for a tuner/amplifier does make a lot of sense. Sound quality can be just as high as with true separates; there is usually a cost saving of between 10 and 20 per cent; a tuner/amplifier takes up a bit less space; there are one or two fewer connections to worry about.

There are a couple of disadvantages, though. You might want a powerful amplifier, but only a modest sort of tuner – or vice versa. Yet a tuner/amplifier with a powerful amplifier section usually has a sophisticated tuner too – so you might save money by buying the appropriate separates. If the tuner section breaks down, and you send the receiver off to be fixed, you lose the use of the amplifier section (and so the rest of your hi-fi too).

In theory, a good-quality music centre should be an excellent way to buy hi-fi: it enables you to enjoy music with the minimum of fuss. But it is very restricting, and if you can get interested enough in hi-fi (reading this book would do fine) to contemplate buying separates, you would probably end up with much better value for money.

Rack systems seem to have few benefits over buying separates, and many disadvantages. Buy separates if you can, or a music centre if you really want a combination. (If it is the appearance of a rack system that appeals to you, search for a music centre that is designed to look like a rack.)

Separates are a sensible way to buy hi-fi, even for the non-enthusiast. But you can reduce the number of bits and pieces you have to buy and match if you get a tuner/amplifier.

What about digital?

You may have heard the word *digital* used to describe discs, or, more widely, to herald the next hi-fi revolution. Many record companies already use digital equipment for making the tape recordings from which they will eventually press analogue discs. These ordinary discs are often marked in some way as digital – but they are ordinary records none the less. Some offer very good sound quality (although nowhere near the quality of a fully digital disc), but this may be as much because of the care taken with the microphone placing, the recording and the disc cutting as with the use of digital rather than analogue tape recorders. Page 22 explains some of the technical background to digital recording.

It seems unlikely that there will be many (if any) competing systems of digital disc and player for the

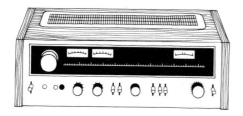

Tuner/amplifier

general public, unlike the video recorder market – but it does seem likely that it will be some time before digital equipment is widely available, and even then the conventional analogue discs and players will be available for quite a few years. So there is no need to worry that your hi-fi will be obsolete overnight. Nor need you be inhibited about buying new analogue equipment now; if nothing else, you will need something to play your existing library of records on. On the other hand, if you like buying new gadgets as soon as they appear, it may be sensible to spend only the minimum necessary on a record deck and amplifier, so you will have some money saved to invest in digital gear.

Choosing brands

It is probably impossible to provide perfect audio reproduction – sound quality so good that even the most expert ears could not tell it from the original sound. But it is possible to get very close – so close in fact that the differences are irrelevant to most of us. If you think that hi-fi equipment is a means to an end – a way of enjoying reproduced music in your own home – then take no notice when buying equipment of those people who counsel you to look for near-perfection: all you need is to find something that will reproduce music just as well as you want it to, and for a reasonable price.

But how *do* you judge quality? There are various directions you may look to for guidance. They all have very serious limitations. Let's look at them in turn.

Standards. You may come across hi-fi equipment marked with the various Standards set for it. Probably the most famous is DIN – the German standards organisation. For equipment to be marked with such a Standard, it must have passed certain tests of technical quality. But these Standards are so low that most halfway to decent equipment should pass the tests easily. And the new British Standards, though often tougher, are still not rigorous enough to be a real buying guide.

Technical tests. You will find references to technical tests in many places. This book provides a guide to them.

But a useful lesson at this stage is that technical tests are often of little help in telling you what hi-fi actually sounds like. Listening tests are often more sensible (but *Listening tests* and *Where to buy* on pages 14 and 15 point out the practical limitations of these).

Manufacturers' specifications. Given our warnings about test data in general, it will come as no surprise that figures provided by manufacturers – often reams of them in glossy brochures or ads – are of little use. They are anyway quoted in different forms by different manufacturers, so not always comparable; they often have so little information on how the tests were made that the results are meaningless; or the test results you need are not given. So, coupled with the fact that the results may well not tell you how good the equipment sounds, a second useful lesson is that such specifications are rarely any help in choosing hi-fi equipment.

Which? magazine hi-fi reports are about the best published information – as far as they go. Our magazine is rigidly independent; samples for testing are bought anonymously in shops, not accepted from manufacturers; testing is highly sophisticated and carefully controlled; all results checked; the reports produced comprehensibly. But hi-fi is only one area among scores covered by *Which?* Inevitably, it can never hope to keep abreast of the sheer numbers of brands and models on the market. If a report coincides with the buying decision you're making, and the models you're interested in are covered, you can rely on it; otherwise, you're out of luck.

Hi-fi choice is good too. Their series of guides – each covering a different type of hi-fi equipment – appear at irregular intervals. The test methods are good and comparative – though they do generally accept samples from manufacturers, and contain advertising. Their reports rely rather too heavily on technical language, and the buying advice is usually not as clear as in *Which?*

Hi-fi magazines generally can certainly be more up to date than *Which?* or *Hi-fi choice*. But it is often difficult to compare test results, as the equipment is usually

Listening tests

One recurring theme throughout this book is that the best way of finding out how a hi-fi system will sound is to listen to it. But critical listening is not easy, and there are many pitfalls which could trap you. Here are some guidelines.

Obviously, you will need some **programme material** to listen to. Hi-fi equipment is now often of a higher quality than the records, tapes and broadcasts played on it – so if you want an accurate idea of what your hi-fi sounds like, you will need to use the very best programme material. If you are used to the sound of live music, then recordings of that will be good material for you to use – because you will have a reference against which you can judge the accuracy of the hi-fi. With music that relies heavily on electronic effects – rock and some modern classical, for example – it is more difficult to establish what is accurate.

Whatever type of music you intend to listen to, though, be sure to include passages which will **show up any defect** in the equipment – slow piano or woodwind for wow and flutter; quiet passages for hiss, hum and rumble; loud passages for distortion; passages with plenty of high frequencies (eg brass) and low frequencies to check on frequency response.

Which? finds it best to use a method of listening known as **A/B comparison**: each item or system is listened to, in turn, comparing it against a reference item or system. The listeners are able to switch back and forth from the reference to the test item. Shops often have A/B switching boxes, but for home use you would have to build one. This may not be easy; a switching box can introduce its own distortions.

Of course, if you want to hear the effect of changing only **one piece of equipment** in a system (the amplifier, say), then the rest of the equipment – loudspeakers and record deck – should be the same for both the test and reference sections.

Once you are in the realms of switching boxes and comparative trials, you should take elaborate precautions to ensure that any **differences you hear are real** ones. For example, take great care that the sound levels between the test and reference items are the same. Check that all plugs and sockets are secure, that any record or tape decks are properly set up and aligned, and that recording heads and pick-up cartridges are clean. If you notice hum, don't immediately blame the hi-fi: the problem could be in the connections – see pages 16 and 17.

It is very easy to fool yourself when carrying out listening tests – for example, by persuading yourself that the more expensive item of equipment really does sound better; or even that you can hear differences in sound quality when really you can't. For this reason, it is best, if you can, to do what are called **blind tests**, so that you don't know which equipment you are listening to at any time.

Which? goes one step further in its listening tests: the panel listens, unknowingly, to the same piece of equipment more than once during the listening trial, as a check on consistency. Even sneakier (but a useful check) is to arrange an A/B test in which occasionally there is no difference between the test and reference systems.

After all this checking, you may find a difference between the test and reference systems that you are certain is real. But you mustn't stop here: you should **check that the difference is inherent in the equipment** and not, say, because the two sets of loudspeakers you were listening to were at slightly different positions in the room (the check here is to swap the two sets of loudspeakers around and listen again). Continue these sorts of checks – some points to watch out for are covered in later chapters – until you are sure that the difference cannot be explained away.

A final warning. Even now, all you really know is that one piece of equipment sounds *different* from another. Do not be fooled into thinking this necessarily makes it sound *better*. Sometimes there will be no doubt, but if you remain unsure about which is the better, even after endless listening and using many different sorts of material, then the chances are that the differences are not worth worrying over. Choose the item that gives you the features you want at the lowest price; spend the money you've saved on your favourite records. Forget about listening tests; relax and listen to the music, not the equipment.

reviewed by different individuals. Comparative tests on the scale of *Which?* are rare. You may, of course, find the mixture of letters, advice, ads, product news etc in these magazines interesting.

Dealers are often very free with advice. Just as the equipment they stock is limited in range, so will be the advice – and usually aimed at persuading you to buy the equipment they do stock. And it would be unwise to rely on their guidance, even about the equipment they stock. So you can largely discount dealers' advice.

Buy on price? You will be told that you get what you pay for. This is no more true of hi-fi than it is of anything else. Obviously there is occasionally a relationship between price and quality. But, more often, *Which?* finds little or no difference in performance – and the cheaper equipment becomes better value for money. Often more money will buy more features – many of them of little real benefit to you. So you don't get what you pay for.

Assume they're all the same? This is the other extreme – and has a little more to commend it. Certainly, if you want to spend, say, £100 on an amplifier, a pin might well be a fairly sensible tool for deciding with. Even in side-by-side tests, you'd be unlikely to spot many differences – especially now the standards of performance are so high, generally. But there remain some differences – and you need to take into account the various features on offer, which do have a bearing on price. So you need something more than a pin.

All is not hopeless, despite the fact you can't rely on Standards, tests, manufacturers, magazines, guides, dealers, or old consumers' tales. With a good, commonsense grounding in hi-fi, you can find your way quite happily to perfectly good buys, at £50 or £1,050. This book should give you that grounding. One of its main jobs is to help you select from those other sources of information the bits that are useful to you.

Where to buy

Many hi-fi shops give demonstrations of equipment. This is an important service: as you will see through the rest of the book, if you are aiming to get the best combination of different pieces of hi-fi equipment, it is essential to listen before you buy. *Which?* looked at how good shop's demonstrations were for its report, *Putting together a hi-fi system*, in June 1979. The results were not impressive. Few shops had a proper listening room, with acoustics like those you would find in a domestic living room, and the loudspeakers properly positioned (which is vital); few shops had many systems on demonstration or indeed had a good range of brands available for sale; although demonstrators were helpful, disappointingly few were at all knowledgeable.

To some extent, how good a demonstration you get may depend on you. Be clear about the amount of money you want to spend; the type of music you want to listen to; what you are buying a system for (background music, serious listening or whatever); and how important sound quality is for you. It is very wise to take some good records with you – and have a healthy suspicion of any of the demonstrator's material until you are sure that it is not going to hide problems rather than expose them.

Home trials, particularly if you get the opportunity to compare different pieces of equipment, are a valuable type of demonstration. Some shops sell equipment on approval – this can be nearly as good, so long as you can either get your money back, or you know there is something else in the store you want instead.

There is no hard and fast rule about the best type of shop to go to for demonstrations. Specialist hi-fi shops are often, but not always, among the best places, and tend to charge about 10 to 15 per cent more than the cheapest places. Discount stores are often cheap, and may sell equipment on approval. Expect to have to work hard if you want to find really good listening facilities.

How much to spend

It is not difficult to spend more on a hi-fi system than on a small family car – indeed it is not impossible to spend more than on a small family house. But, comfortingly, even good-quality hi-fi can be relatively reasonable in price.

CHAPTER 1

Giving advice on how much to spend, and how to apportion the money, is very difficult though – prices in the hi-fi business vary wildly both up and down. As a very rough guide, £300 should buy a reasonable-quality system of record deck and cartridge, tuner/amplifier and speakers; £500 should give you very good sound. You would have to be a very critical listener to justify spending more than that on the basic system. *Which?* advice is that loudspeakers make the greatest difference to the sound of a system, so do not be afraid to spend quite a lot on them if you have to – say a third or more of the total outlay on a system costing between £300 and £500.

Upgrading

Upgrading is changing part or all of your hi-fi system for something that you think and hope is going to be better. For many hi-fi addicts, it is a continuous operation, and one that can prove to be very costly. One of the messages of this book is that this path of continuous upgrading is one you need never step out on: even relatively inexpensive equipment, if chosen wisely, can give sound quality as good as most people will ever need. But if you are unhappy with the sound of your hi-fi, you will want to know how to go about improving matters. Here are some guidelines, but of course you must refer to the relevant chapters of this book for the details.

● If it is only a radio tuner or tape deck that you are unhappy with, the thing to do is to replace it. Great strides have been made in these areas even in the last two or three years, and modern machines may well perform much better than the ones you have. With a cassette tape deck, the problem may be that you are using the wrong tape for the machine; it is worth investigating this before buying a new machine.

● Whether to replace a record deck that you don't like the sound of is a little more complicated. If the problems are mechanical – such as rumbling noises, or wobbly tones – then the deck should be replaced; modern record decks should be free from these problems. If other aspects bother you, then a change of pick-up cartridge may help – though in choosing one, it is fairly important to make sure it will match the record deck, the amplifier input, and the loudspeakers. Simply altering the position of the equipment in your room could help, too.

● If you are generally unhappy with the sound quality of your set-up, and the same types of fault are audible whatever the source you are listening to, then the chances are that your speakers or amplifier are not up to scratch. Our advice is to suspect the speakers first. Amplifiers are likely to be the source of trouble only if relatively elderly – very roughly, more than five years old for most brands; if underpowered for your speakers; or if just plain faulty.

Hi-fi repairs – d-i-y?

Which? surveys confirm that hi-fi equipment is pretty reliable. And servicing and repair seem to present little problem.

But if audio equipment isn't working properly, it's worth making some basic checks before you assume there's something seriously wrong with it; you could save yourself unnecessary labour charges.

To start with, don't make random adjustments in the hope of improving matters if your equipment has previously been working satisfactorily. If you *do* attempt adjustments:

● stick to the ones mentioned in the manufacturer's instructions, or below; for example, don't oil moving parts unless the instructions say you can

● take careful note of the existing set-up before making any alteration – then you can at least put it back if there's no improvement

● unless you're qualified to do so, don't remove covers while the mains supply is connected; don't use spare parts other than those supplied by the manufacturer.

Complete failure. If nothing works, including any dial lamps or indicators, check that the mains plug is wired correctly and that the plug fuse hasn't blown. If the fuse has blown, try replacing it with another of the correct rating (3-amp), *once only*; if it blows again, the chances are that there's a fault which will need expert attention. There are often other accessible fuses at the back of hi-fi equipment – consult instructions.

Crackles, hum or no sound on one or both stereo channels are often caused by loose or faulty connecting leads and plugs between pieces of hi-fi equipment. Try gently moving each plug in its socket (keeping the volume turned fairly low) to pinpoint the fault. If only one channel has fault symptoms, you can often locate the trouble by swapping over leads, left to right and right to left (switch off while you make the changes). If you do find a fault, you'll have to buy a new lead if you can't cope with the rather fiddly soldering involved.

Hum is worth special attention. Unplug all connecting leads, except the loudspeaker cables. If you still have hum, then there is a fault in the amplifier or the amplifier isn't properly earthed to the mains supply (or you need a better quality amp). Otherwise, connect up and switch on each component in turn. When you've found the guilty one, check if you can reduce the hum by moving or turning the component away from the amplifier, or by moving the connecting leads away from the mains or loudspeaker leads.

If there is still hum from a record deck, which reduces when you touch the pick-up arm, it might be that the arm isn't earthed; check by holding a piece of wire from a bare earthed screw on the amplifier to the arm. If this helps, and there are earth tags on the amplifier and record deck, connect a lead between them. If this doesn't sort it out, or if there are no earth tags, you may need professional help in deciding what to earth to what.

You can get hum from too many earth connections, rather than too few – which may give rise to earth-loops. The best plan for minimum hum is to disconnect any earth wires in the mains leads of all components except the amplifier and to rely on the connecting leads to provide the earth paths for these. But this isn't the best plan for maximum electrical safety; instead, try disconnecting the outer braiding of the leads between the guilty component and the amplifier at one end. Using leads with different screening arrangements might also help.

(You can't, of course, suffer from this sort of earth-loop if the components have only a two-core mains lead, as many do these days.)

Earthing the pick-up arm and record deck separately from the earths on the pick-up lead to the amplifier might also cause hum-loops.

If the pick-up cartridge has a metal body, you might improve things by insulating it from the pick-up arm using nylon rather than metal spacers and screws.

On an amplifier, there may be *loudspeaker fuses* which will blow if the wires to a speaker accidentally touch each other.

Distortion on records can be caused by accumulated dirt on the stylus, or by a damaged stylus. See *Record and stylus care*, page 78, for details.

Cassette deck problems. Muffled sound, ineffective erasing, and unsteady tape speed can all be caused by an accumulation of particles shed from the tape. Regular cleaning of the tape heads and adjacent parts is essential. See *Tape deck care*, page 110, for details.

Distortion, crackles and other noises on VHF radio are often not actual faults, but interference caused by cars or domestic appliances. This can often be cured by a better aerial; if you've got a good outside aerial already, it may have been shifted by high winds. Leakage of rain water into the connecting cable is another possibility.

More complex d-i-y repairs. It would probably be unwise to attempt substantial d-i-y work inside the guarantee period in case you invalidate it. In general, don't do anything you are not confident about; you may anyway have difficulty getting hold of parts, or even service manuals.

Chapter 2 All about sound

Hi-fi is (or should be) only a means to an end – a way of hearing music or speech in more or less the same way that you would be able to hear at, say, a live concert. Hi-fi is only a medium – and arguably the best hi-fi system would be totally unobtrusive, neither adding anything to, nor taking anything away from, the sounds as they pass through the hi-fi chain. So, to understand hi-fi performance, it helps to know how musical sounds are made up. That is what this chapter is about.

What is sound?

The first answer to this question is that 'it all depends on what you mean by sound'. Ancient Greek philosophers mused on whether or not a tree falling over in a deserted forest makes a sound. A modern answer would be: if by sound we mean the vibrations produced in the air by the tree falling over, then indeed the tree makes a sound. But if by sound we mean the response of our hearing faculty – the ear and the brain – then there is no sound.

The distinction is important if only because the whole of the stereo illusion relies upon it: the distribution of sound coming out of the two loudspeakers is nothing like the distribution in the original concert, but the hearing faculty can be persuaded into thinking that it is.

Sound waves

Sound starts with a vibration – a hammer hitting a block of wood; an opera singer's vocal cords vibrating; a string being plucked or scraped.

For these vibrations to reach the ear they have to be transmitted from the source of the vibration in some way. In air – the most usual medium for transmitting sound – are gas molecules (the very very minute particles that make up the substance of the gas), constantly darting about the place, cannoning off each other, and flying off in new, random directions.

The molecules that happen to be around a vibrating violin string will be affected by the vibrations, alternately being pushed towards each other (thus increasing air pressure) and then pulled apart (lowering pressure). So the air immediately around the plucked string is alternately of slightly high pressure, then slightly low; this cycle continues as long as the string vibrates.

The air surrounding the air surrounding the string is then affected by this pushing and pulling of the molecules, and in turn, this region of air is subject to repeated high/low pressure variations. And so it goes on, with the pressure variations rippling out in all directions. This rippling is called a **sound wave**.

Once the sound waves have reached the ear, our second definition of sound comes in. The vibrating molecules set up sympathetic vibrations in the mechanisms of the ear, and the brain processes the vibrations, producing the sensations that we call sound. (Exactly how the brain does the processing is not clear, but – interestingly for today's hi-fi – the process appears to be *digital* rather than *analogue*; see page 22.)

This simplified sound wave shown opposite has two characteristics – its *frequency* (how quickly the wave pattern repeats) and its intensity or *amplitude* (how big the wave is). When the ears pick up the sound wave vibration, the brain gives it similar characteristics. But the brain does not hear quite what physicists say the wave is doing. Instead, it perceives frequency as *pitch* and amplitude as *loudness*. There are important differences between frequency and pitch, and between amplitude and loudness.

Frequency and pitch

One of the major characteristics of sound waves is that they repeat themselves. The wave opposite repeats itself every one-thousandth of a second (put another way, its frequency is 1,000 repetitions per second). Repetitions per second is a cumbersome phrase and physicists have shortened it to Hertz – pronounced hurts, and shortened still further to *Hz*.

The difference we hear between sound waves of frequencies, say, 1,000Hz and 250Hz is the difference we describe as **pitch**; the higher-frequency sound has the higher pitch. Each note on a musical stave has its own pitch corresponding more or less exactly to a particular frequency, as the diagram on page 21 shows.

The pitch we hear is almost directly related to the transmitted frequency – though for pure tones at frequencies of around 200Hz to 2,000Hz, the pitch of a steady frequency appears to *rise* slightly as the tone decreases in volume.

The range of frequencies that a human ear can hear depends mainly on the age of the listener. Young children should be able to hear notes between about 20Hz and 20,000Hz. High-frequency hearing drops off with age – perhaps down to 5,000Hz or so.

FROM VIBRATIONS INTO WAVES

The vibrating string causes the molecules of the air (represented by a mark) to be pushed together: the closely-spaced lines represent this area of higher pressure.

A split-second later the higher pressure region has moved outwards from the vibrating string as the string pulls the molecules near it apart causing a region of lower-than-normal pressure.

The high pressure region therefore moves out from the string, being replaced as it moves by a region of low pressure which in turn is replaced by another region of high pressure and so on. To an observer just looking at point A, the pressure seems to vary in a cycle from normal to high to normal to low and back to normal again, time after time. To an observer of the whole sound wave, it looks as though a single ripple of pressure is moving outwards from the string.

In fact, the air molecules themselves don't specially move in the direction of the advancing wave – they continue their random course, and simply pass on the push or pull that they have experienced to adjacent molecules. When the advancing wave reaches the ear, it sets up vibrations in it. (The wave continues beyond the ear, and the string causes similar sound pressure waves in all directions.)

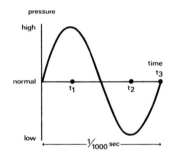

It is easier to see that alternating pressure variations set up a wave if the marks representing the pressure variations in the vibration diagrams on this page are redrawn. Before the wave has reached the point A the pressure is normal. At a slightly later time – call it t_1 – the pressure is high, and at a yet later time (t_2) the pressure is low. At time t_3 the air is back where it started – at normal pressure, with the pressure increasing. If these, and many other points are marked on a graph, the result will be a wave-shape as above. The vibrations can vary a lot in rapidity: in this example, the time from start to t_3 is just one-thousandth of a second.

Amplitude and loudness

The bigger the sound wave is (ie the larger its amplitude), the more **power** that sound has. It is this power that registers with our ears and brains as loudness. Like all power, sound power can be measured in watts, though the numbers involved are very small: the amount of acoustic power per square metre for the very loudest sound the ears can tolerate is only about one watt.

The loudness we hear does not increase as dramatically as the power behind the sound. For a sound to *double* in loudness, the power in the sound wave has to increase by roughly *ten times*.

Decibels. To avoid huge and tiny numbers, and to ensure that sound power is expressed in units that bear some relationship to how loud the sound appears, sound power is usually measured in terms of **decibels** (written **dB**). Put simply, every tenfold increase in power (or, roughly, every doubling in loudness) is equivalent to an increase of ten decibels. For example, if we arbitrarily call a power of one-hundredth of a watt 100dB, then a tenth of a watt is 110dB, and one watt 120dB.

Decibels are important throughout the whole of hi-fi. One important point to note is that a decibel is not a number, but is a *ratio* between two numbers. Decibels

PITCH AND LOUDNESS

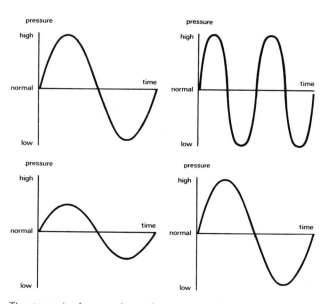

The *top* pair of waves have the same amplitude. But there are more repetitions a second in the *right-hand* one than in the *left-hand* one – it has a higher **frequency**. Another way of looking at this: its **wavelength** is shorter.

The *bottom* pair have the same frequencies. But the *left-hand* one has smaller pressure variation or **amplitude** (the resulting sound would be quieter).

PHASE

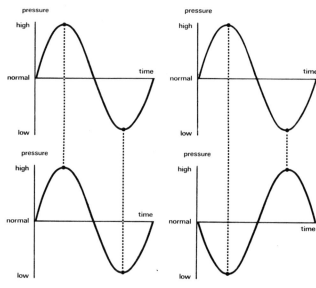

The two waves on the *left* move up and down exactly in step: when waves move like this, they are said to be **in phase**. Whenever waves are not exactly in step – however close or far apart they are – they are **out of phase**. The waves on the *right* show the special case when one wave has reached its highest point at the time when the other wave is at its lowest point – these waves are **180° out of phase**.

are, however, widely used as though they were actual numbers. This is particularly unhelpful with manufacturers' quoted equipment specifications. Unless you know what the reference quantity is – the other part of the ratio – the dB rating is useless to you.

The reference level in sound measurements is the threshold of hearing – the quietest sound a person could hear. This is given the level of zero dB: a sound ten times this power is therefore at 10dB; a sound one *million million* times this power is still only at 120dB and represents the threshold of pain. Higher-powered sounds don't sound any louder; they just hurt. As a very rough rule of thumb, the ear is not sensitive to changes in loudness of less than about 3dB.

Perceived loudness is related to frequency as well as sound power. From about 200Hz downwards, and from about 5,000Hz upwards, the ear becomes less sensitive to sound – so to keep the apparent loudness the same at low and high frequencies as at the middle frequencies, the sound power would have to increase.

The difference between the quietest and loudest sounds that an orchestra plays – its **dynamic range** – can depend on the type of music. It might be as low as 40dB for music of Mozart's day, and as much as 70dB or more for a Mahler symphony. The orchestra clearly doesn't want its quiet, sensitive passages to be lost in the noise from the audience, or the air conditioning in the concert hall. This sort of noise might be at about 35dB, so it will play a *ppp* (very, very quiet) at about 40dB. And if the orchestra has a dynamic range of about 70dB, then its *fff* (very, very loud) will be at about 40 + 70 = 110dB.

Decibels are used to describe ratios of two *voltages*, as well as two powers. So the dB is equally useful in the electronics side of hi-fi – ratios like signal to noise, the gain of an amplifier, relative boosting of different parts of a frequency response.

Real music

Sound an A on a piano and the same note on, say, a violin. Even ignoring the difference in the way in which the two sounds start, it is very easy to distinguish the two instruments: their *tone colour* is totally different, despite the fact that they are both playing the same note.

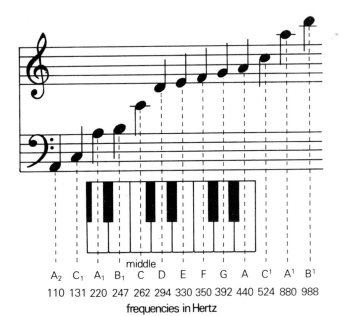

Some notes on the musical scale, together with their frequencies in Hertz. The frequencies are not evenly spaced, but the distance between successive notes gets greater as the pitch increases. In fact, *the frequency distance of one octave doubles in each successive octave.* Standard pitch is nowadays based on the A above middle C, that is a frequency of 440Hz. As octaves are notes in the ratio 2:1, the next A up (A^1) will be at 880Hz, and the next one down (A_1) will be at 220Hz.

The notes, of course, could represent any frequency, providing the 2:1 relationship between octaves was maintained. As recently as the early 1900s, A could be up to 456Hz, and books on sound often define notes so that middle C ends up at a frequency of 256Hz, rather than 262Hz.

Mathematically, it is quite easy to find frequencies for the twelve semitones within an octave in such a way that the frequency ratio between any two semitones is always a constant – 1.059:1. Most fixed-note instruments such as the piano adopt this *equal-temperament tuning* – and it is because of this method of tuning that Bach was able to write his *Well-tempered Clavier* preludes and fugues. The problem is that this makes the intervals between some notes in an octave rather awkward. For example, a major third becomes a ratio of 5.03:4. To keen ears, this is a significantly less pleasant interval than the ratio of a true major third (5:4).

CHAPTER 2

HOW LOUD?

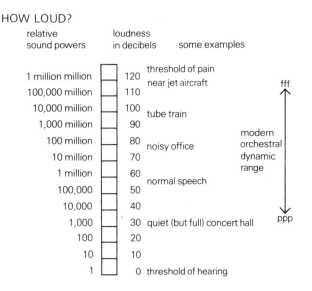

Why? Because, with very few exceptions, no instrument gives out a single pure frequency even when only one note is being played. Instead, as well as vibrating at a *fundamental frequency*, the air is made to vibrate at multiples – two, three, four times and so on – of the fundamental frequency. The fundamental is called the *first harmonic*; the frequency at two times the fundamental is called the *second harmonic*; that at three times the *third harmonic* and so on. The relative intensity of these harmonics varies from instrument to instrument – and it is this difference in the strength and number of the harmonics that is largely responsible for each instrument's unique sound.

Again, what the ear hears is different from what the acoustic expert can see. It is easy in the laboratory to separate and analyse the different harmonics in a complex sound. The brain, it seems, is specifically built not to do this – many people, indeed, would find it difficult to

---

Digital recording

The kind of audio reproduction discussed in this book is **analogue**: the electrical signals passing through the equipment are direct representations of (analogous to) the sound waves creating them. Although this is the most obvious method of reproduction, it is not the only one. Another method involves measuring the two features of a sound wave that are important – its frequency and amplitude – and expressing these as a number. This, along with countless thousands of other numbers each representing other parts of the sound wave, can be stored and transmitted electronically. This is a **digital** method of transmitting information.

In practice, what happens is that at fixed, small intervals of time, the sound wave is measured (in the jargon, *sampled*). This gives the amplitude of the wave; because the measurement is repeated at regular intervals, the frequency of the sound wave is known too. In effect, the sampling rate sets the highest frequency of the system (audio frequencies of up to 20kHz are no problem) and the accuracy of amplitude measurement sets the signal to noise ratio (easily 80dB, or more, even without any noise reduction).

The rate of digital numbers produced by the sampling is too high to be stored in a conventional audio tape or disc – but it can be stored relatively easily on video tape or video disc. Even if you were keen to try this, before you can hear the tape or disc, the digital stream has to be reconverted to an analogue one; as yet, there are no digital loudspeakers you can go out and buy.

It seems like a lot of trouble to go to – converting analogue sound to digital, capturing it on an expensive video deck, and then reconverting it to analogue. So what are the advantages? Sound quality can be much higher: wow, flutter, noise and distortion either do not exist in digital form or can easily be corrected electronically. And there need be no degradation of the sound quality during the recording and re-recording stages of a disc. The disc you buy should be capable of exactly the same sound quality as the original recording of the sound in the music studio.

ALL ABOUT SOUND

separate even two notes of different fundamental frequencies (particularly if they were, say, an octave apart) let alone sort out the harmonics within one note.

Attack and decay. Real music does not consist of constant notes, but of notes that stop and start. With some instruments, the starting (**attack**) or stopping (**decay**) happens relatively slowly; with others, they happen very quickly. An instrument's attack and decay is an important part of its overall sound quality – as important as the arrangement of the harmonics; indeed, many instruments (particularly percussion) are nearly all attack and decay, with little in the middle.

Audio equipment finds fast stopping and starting of notes – called *transients* – more difficult to cope with than steady tones. Some technical tests (particularly of amplifiers) are specially designed to check the transient behaviour of equipment.

REAL MUSIC WAVES

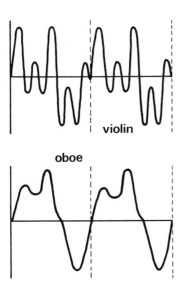

The diagrams on the *left* show the form of a wave for a violin and an oboe (actually, an oboe organ pipe). The difference in the two shapes is one of the reasons why the two instruments sound different.

The diagrams on the *right* show how the oboe waveform is made up mainly from a fundamental frequency (*1*) and just the second harmonic (*2*). The frequency of the fundamental gives the completed tone its pitch; the relative strength of the harmonics gives the tone its characteristic oboe sound. Adding together the distances from the centre lines to each of the waves at various points along the wave – noting that distances above the centre line count as positive, and those below count as negative – gives a combined wave (*1+ 2*) which looks roughly like the oboe wave form.

Note that the waves which go to make up the completed waveform have a very simple, regular shape – the same shape as the rest of the waves drawn in this chapter. This sort of shape is called a sine wave. A simple, single sine wave usually at a frequency of 333Hz, 400Hz or 1,000Hz – the keyboard on page 21 shows the nearest musical notes to these frequencies – is widely used for testing hi-fi equipment. See next chapter.

Chapter 3 Hi-fi fundamentals

This chapter serves mainly as an introduction to the commonest performance features of hi-fi equipment – frequency response, dynamic range and signal-to-noise ratio, distortion, wow and flutter. These are issues that crop up throughout the book: distortion, for example, is as much of a problem with cassette decks as with amplifiers. Dealing with them here saves repetition and establishes the ground rules the book follows in helping you interpret such data.

But first, the link between electricity and hi-fi; some information about graphs and their use in this book; plus a guide to units and measurement.

What has electricity to do with hi-fi?

The essence of hi-fi is that it uses electricity, electronics and magnetism to store, transmit and reproduce sound. It is not essential to understand anything about these processes before you can set about buying and using hi-fi equipment – any more than it is necessary to know anything about mechanical engineering before you can drive a car. But, just as some understanding about why a car pulls better and quicker in a low gear can help make you a better driver, so a basic knowledge of electricity and sound can help you appreciate better what hi-fi equipment can do.

An electric **current** is set up by the movement of electrons in a material. The rate of movement, and therefore the amount of current, depends on the electrical pressure between the ends of the material, and on the resistance the material puts up to electron flow. The greater the pressure, and the less the resistance, the bigger the current.

Current is measured in **amperes** (amps, or A, for short) and the electrical pressure, or **voltage**, in **volts** (V). The resistance the material puts up is called its electrical **impedance**, and is measured in **ohms** (Ω). Most materials either offer so little resistance that current flows virtually unhindered – *conductors*; or they put up so much resistance that effectively no current flows at all – *insulators*. Electrical components that put up known amounts of resistance between these two extremes are called **resistors**; their impedance can easily be measured.

There is a strict relationship between voltage, current and impedance: the current (in amps) through a material is equal to the voltage (in volts) divided by the impedance (in ohms). So if you know any two of the quantities, you can calculate the third.

Electricity's ability to do work depends on **power**. Power is measured in **watts**; again, there is a relationship between this and the voltage and current: power is equal to voltage multiplied by current.

In fact, electricity flows steadily in one direction like this only if it is **direct current** (dc). Mains electricity is almost all **alternating current** (ac); strength and direction vary. (This makes it easier to transmit over long distances.) A graph of how the current varies with time would look like one of those on page 20, with the amplitude of the sound becoming the size of the electric current. The frequency of such an electric wave would be (in the UK) 50Hz, and the voltage 240V.

So, if it is possible to represent one type of sound wave by one type of alternating current electrical signal, is it possible to represent all types of sound wave in electrical terms? The answer is yes – within very close limits. Indeed, almost all of hi-fi is about how well and accurately you can do this – and how far such accuracy matters.

Since the current and voltage in an ac signal is constantly changing, the relationships between impedance and power have to be redefined slightly. A useful way of describing alternating currents and voltages is to give them the same values as those of the direct current and voltage which would be needed to produce the same power that the alternating ones produce. So, an electric fire drawing 5 amps at 240 volts on ac would produce the same amount of heat as one drawing 5 amps at 240 volts on dc supply. This effective value is called the **root-mean-square** value (**rms**) of the wave. And, unless otherwise stated, you can assume that voltages and current in ac circuits are always given as rms values. (Strictly, what is almost always called **watts rms** is watts average.)

Capacitors and **inductors** are electrical components in the same way as resistors. They have a lot to do with electronic circuits and, like resistors, they can restrict the flow of electric current. Unlike resistors, though, their restricting effect varies with the frequency of the alternating current in the circuit: for capacitors it *decreases* as the frequency increases; for inductors, it *increases* as the frequency increases.

Most of the time, circuits contain bits of capacitance, bits of inductance and bits of resistance – so the current restricting properties are rather complex.

Capacitors and inductors change the phase between voltage and current (which resistors do not). The two will be out of step with each other, by an amount that depends on the relative amounts of capacitance, inductance and resistance. This change of phase can, in some circumstances, mean a loss of sound quality.

Graphs, and how to read them

Many people find a picture of an event easier to understand than its written description; one picture is worth a thousand words. On the other hand, some people have a blind spot about graphs. It's worth persevering with this, as many hi-fi performance measurements are difficult to follow without graphs. A graph is simply a detailed picture showing how one thing varies as you vary something else. For example, a sales graph (top of next column) shows how a company's income goes up and down as the months or weeks roll on. The scale along the bottom of the graph represents time; each additional 10mm is an additional month. The scale up the side represents income; each 10mm is £1,000. Dots are marked in the middle of the graph to show sales income – for example, at the beginning of April sales were £3,000, May £1,000. Since the figures are collected only month-

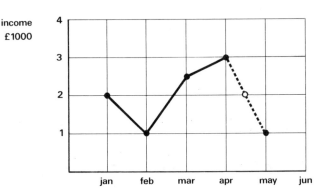

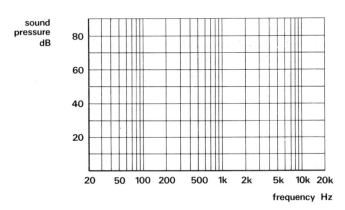

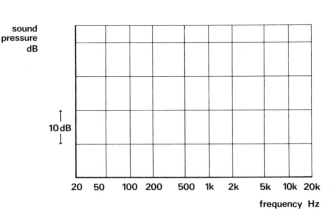

ly, you don't know what sales in, say, mid-April looked like – but you can make a guess by joining up the dots for April and May with a line (the guess is £2,000).

The scales on this graph are *linear*: each unit along a particular scale (or *axis*) represents the same amount. The graphs used in hi-fi work are often not linear: this may be useful to ensure that the graph more nearly corresponds with how the ear hears sounds. For example, the ear is not interested in absolute frequency differences, but in the ratio of one frequency to another. So frequencies are often drawn on what is called a logarithmic scale: each doubling in frequency is represented by a fixed unit along the scale.

The middle drawing opposite shows a piece of typical graph paper used for audio work. Frequency is plotted along the bottom – the *horizontal axis*; the distance between each successive doubling in frequency (ie between each frequency octave) is the same. Similarly, the distance between each tenfold increase – *decade* – is also a fixed distance. The designer has started drawing in marks at equal distances of 10Hz apart from 20Hz up to 100Hz – but by then the lines are getting so close together that, for the next decade (from 100Hz to 1,000Hz) he has to draw them 100Hz apart and so on. This method allows good accuracy in reading and drawing the graph on graph paper, but tends to obscure the fact that the distance between each octave is the same. So the graphs drawn for this book have been simplified, with the frequency marks placed roughly equal distances apart, but with each value being double (or thereabouts) the value of the mark on its left. The bottom drawing opposite shows our method.

In many cases, what we want to know is how the sound output varies with frequency, so the scale up the side of the graph (the *vertical axis*) shows sound output. As the ear reacts in a non-linear way it is useful if this scale is non-linear also. Using decibels (relative to a reference) takes this into account – so the output scale is linear in decibels.

Numbers – little and large

Hi-fi performance is measured in numbers spanning huge ranges – from as little as 0.000 000 000 000 1 (ie one *million millionth*) to numbers that are greater than 1,000,000,000,000 (ie one *million million*). The obvious method of expressing this large range without getting bogged down in zero-counting is to use a set of names to express different sizes.

Rather than adopt the rather random intervals of our Imperial system, the universal hi-fi one is based on the metric principle of deciding on a set name (say *metre* for distance) and adding short words – *prefixes* – to it to describe an amount ten, a hundred, a thousand; or a tenth, a hundredth, a thousandth times the basic unit. So one hundredth of a metre is a *centimetre*; one-thousandth a *millimetre*; one thousand times a *kilometre*. The same prefixes can be used whatever is being measured: a *milliwatt* is one-thousandth of a watt; a *millivolt* is one-thousandth of a volt. Since the units themselves are often written as symbols, it is useful to be able to write prefixes as symbols, too; the chart shows the values of the prefixes, and the symbols used. Don't get m and M confused (they often are, in written material).

G	giga	1,000,000,000 times (one thousand million, sometimes written as 10^9)
M	mega	1,000,000 (one million, 10^6)
k	kilo	1,000 (one thousand, 10^3)
m	milli	0.0001 (one thousandth, 10^{-3})
μ	micro	0.0000001 (one millionth, 10^{-6})
n	nano	0.0000000001 (one thousand millionth, 10^{-9})
p	pico	0.0000000000001 (one million millionth, 10^{-12})

Frequency response

For reproduced music to sound natural, one important requirement is that it should have the same *tonal characteristic* as the original. This means that hi-fi equipment must treat each frequency passing through it equally – discriminating neither for nor against any particular frequency or range of frequencies. If the equipment gave a boost to some low frequencies, for example, instruments playing those notes would stand out unnaturally, perhaps giving the music a 'boomy' character. If it made very high frequencies quieter than they should be, then the harmonic structure of instruments would change, at best making the reproduction dull-sounding, perhaps changing the sound of the instruments themselves. So the frequency response of a piece of hi-fi equipment is one measure of how well it preserves the tonal characteristic of the music.

Making the measurement. The usual method of test is to put in a signal at the input of the equipment, changing the signal's frequency from very low (20Hz, often) through to very high (20kHz), and measuring the amplitude of the signal at the output, plotting the results on a graph. The input signal has a constant amplitude, whatever its frequency – so if the equipment treats all frequencies the same then the output will be a constant amplitude at all frequencies too; any deviation shows that the equipment is less than perfect.

When plotted on graph paper the output signal, like the input signal, will be a straight horizontal line, if the equipment is faultless. A useful shorthand for this condition is to say that the equipment has a *flat frequency response*.

FREQUENCY RESPONSES – AMPLIFIER...

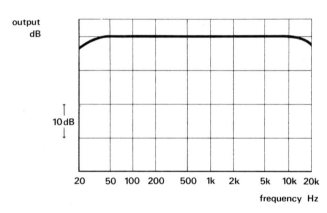

PICK-UP CARTRIDGE...

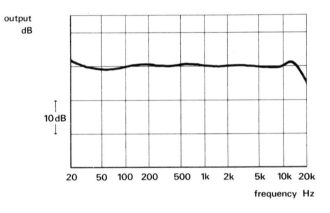

These graphs show some frequency response plots for different types of equipment.

For an **amplifier** (*first graph*), the response is usually ruler-straight over a wide range of frequencies, with gentle rounding at each end: in the jargon, the response *falls off* at low and high frequencies. It is not easy to decide exactly where the rounding starts, and in any case, a small deviation from the flat may not produce noticeable changes in the character of the music. So the extent of a frequency response is usually measured at the points where the response has changed by some percentage below or above some nominal level. Since the scale is already in dB, the percentage is simply a fixed difference in dB: one level often used is a change of 3dB from the level of the signal at 1kHz. In this graph, there is no increase in output over the level at 1kHz, so the response limits will be the frequencies at which the output has dropped by 3dB – in the jargon, the *minus 3dB points*. In this example, the minus 3dB points are at about 20Hz and 20kHz.

The response for a **turntable pick-up cartridge** (*second graph*) is measured by using a special test record on which is recorded a tone changing in frequency from 20Hz to 20kHz. A perfect cartridge will play back the tone with a flat frequency response; as the graph shows, the response from this cartridge is not all that flat. The problem is that the response can be described in many ways, depending on how you care to define the limits. Very roughly, the wider the response the better, and quoting the results for the – 3dB points from 1kHz would give a result stretching from below 20Hz to 20kHz. This apparently good-looking result arises because most of the deviation from 1kHz is *upwards*. However, quoting the *plus* 3dB points makes little difference: they give a response stretching from 20Hz to over 20kHz (because the peak in response at 12kHz is a little less than 3dB). Perhaps the fairest result to quote is the smallest limits between which the

Dynamic range and signal to noise ratio

Even the quietest audience at a concert makes some noise; if the orchestra is to be heard over that it must play its quietest passages at least as loud as the background noise. Similarly, even the best electronic equipment makes noise, and the smallest amplitude signals must pass through the electronic system at a higher level than this if they are not to be lost.

Depending upon where in the system the noise is being generated, turning up an amplifier volume control can help to make the smallest signals larger than the noise, so that they will not get lost. However, the largest signals (which represent the loudest sounds) might then be too large for the system. Signals that are too large can cause *distortion* (page 32).

The difference between the largest signals that can pass through equipment without causing distortion and the smallest signals that can pass without being lost in the noise is called the *dynamic range*, and is synonymous with the dynamic range of an orchestra discussed on page 21. To reproduce an orchestra having a dynamic range of 70dB, the equipment must have a dynamic range of at least 70dB; and to play back the recording in a room where the noise level (including the contribution made by the equipment running) is 35dB, the loudspeakers will have to provide a maximum undistorted sound level of 105dB. This is no mean feat, and quite a few loudspeakers would be unable to cope (in practice, though, not many recordings will have a dynamic range as wide as 70dB).

In general, the *signal to noise ratio* is the ratio between any level of signal and the noise. But usually it implies the maximum undistorted signal possible; the signal to noise ratio means much the same as dynamic range.

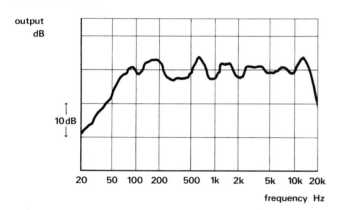

AND LOUDSPEAKER

response changes by plus or minus 3dB from its level at 1kHz the (± *3dB limits*): in this case, about 20Hz to 20kHz.

The moral is: never trust a response quoted simply as, for example, *20Hz to 20kHz* until you know exactly what the definition of those limits is. Better, look at the frequency response graph itself.

The response of a **loudspeaker** is shown in the *third graph*. Similes with mountain ranges are widely used when describing frequency responses in words. The boost in high frequencies in the middle graph is called a *peak*; this graph would be described as very *hilly* compared with that for the amplifier. But it is probable that, in practice, you wouldn't hear the effect of most of the smaller dips and bumps.

The moral here is: even from a graph, it is not easy to predict what the tonal character of a piece of equipment will be.

Making the measurement. There are many ways of making signal to noise ratio measurements, all of which give different results. So you need to know what to look out for.

The first problem is what **noise level** is used. Noise keeps varying in level, usually in a random pattern: quiet one split-second; relatively loud the next. Most methods choose to measure the *average* noise level, which is fine for hiss-like noise (the sort of noise usually encountered in audio equipment). Some methods measure the highest *peaks* of the noise – this would make the signal to noise result look much worse.

Noise can cover all frequencies from low to very high but the ear is more sensitive to the middle-high frequencies (around 1kHz to 5kHz) than it is to the very low ones – so noise often appears as high-frequency hiss. To take this into account in the measurement, it is usual to connect an electrical filter to the meter to boost the middle-high frequencies (and so give them greater emphasis in the meter reading) and cut the low and very high frequencies. This is known as *weighting the reading*, and there are a number of different weighting filters (see the graph, opposite page). Look out for results based on *CCIR curves*: they are less optimistic than those based

HOW MUSIC CAN SUFFER IN REPRODUCTION

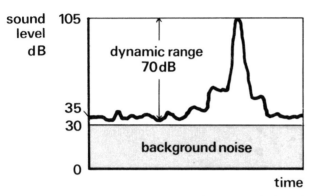

As played in the concert hall, the music starts quietly and grows in volume. Then there is a loud peak – a cymbal crash perhaps – followed by a very quiet passage. The quiet passage is 5dB above the level of the general audience and auditorium noise; the peak is 70dB above this.

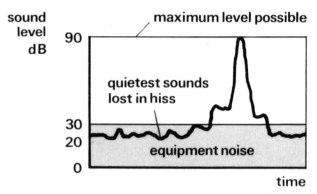

By turning down the volume a bit, the cymbal crash is heard without distortion. But the quiet passages will now be vying with the equipment noise for audibility. (In any case, depending on where in the system is being overloaded, the volume control might not be able to relieve distortion.)

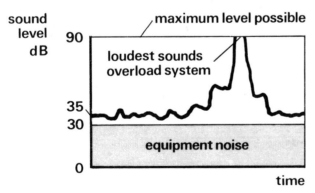

The same passage is now replayed on hi-fi equipment. The quieter parts are reproduced at the same level above the equipment noise as when they were being recorded. But the cymbal crash is too loud for the system and overloads it, causing distortion.

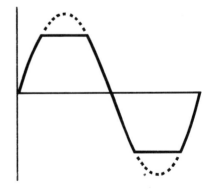

When a sine-wave signal overloads, it has its top and bottom chopped off: the wave changes shape and looks like a square wave which produces a nasty, rasping noise – the sound of the sine wave has been distorted. This is often called in the jargon *driving the signal into clipping*.

on *A or DIN curves*.

The second problem is what **level of signal** to use. Here, the standards used vary widely. One approach is to operate each piece of equipment flat out. This is fine for some components – the result would be the best signal to noise ratio that the equipment was capable of – but would make comparing results for, say, amplifiers of different power ratings, very difficult – because amplifiers are generally not run flat out, but at similar levels whatever their power rating. Here, using a fixed, standard signal level is more sensible.

On the other hand, standard levels are not always a lot of use. For example, some cassette deck signal to noise ratio measurements are referred to *Dolby level* – a fixed, recorded signal level.

WEIGHTING FILTER CURVES

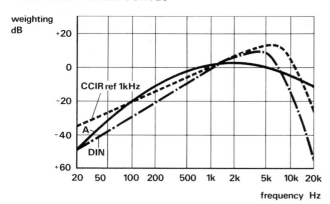

Most laboratories find that the **CCIR curve** gives the best correlation with what listeners think about noise, but many manufacturers use the **A curve** or the **DIN curve** – both of which tend to make the meter reading for noise lower, and so make the signal to noise ratio result look better.

According to the Standards, the CCIR curve which passes through 0dB at 1kHz (abbreviated to *CCIR ref 1kHz*) should be used with a form of peak reading meter. A newer Standard specifies the use of a CCIR curve ref 2kHz (the same curve, but shifted so that it passes through 0dB at 2kHz) and an average reading meter – **CCIR/ARM** for short. Many reviewers, including *Which?* are starting to use this new standard; results will look about 6dB better than CCIR/ARM/ ref 1kHz.

But the difference between Dolby level and the maximum undistorted level you can record will vary depending on the tape and deck used – so the signal to noise ratio of a cassette deck referred to Dolby level tells you nothing about what the overall dynamic range will be.

Finally, with some pieces of equipment, the input and output terminals have to be **loaded** in some way to simulate whatever would be connected across them in practice – pick-up cartridge, loudspeakers etc. Often, this means a complex circuit or, say, a real pick-up cartridge carefully screened from interference and correctly connected to the pick-up input.

To sum up, in order to compare signal to noise ratios *properly* you need to know:
- **how the noise was measured**
- **what signal level the noise is being referred to**
- **how the input and output terminals were loaded.**

Frequency response and signal to noise. There is a link between signal to noise ratio and frequency response. Since noise is most apparent at high frequencies, a response which does not extend very high will cut noise and so give a better signal to noise ratio. This, after all, is what happens when we turn down a treble control or use a high filter.

But sacrificing an extended frequency response for the sake of a better signal to noise ratio is not particularly successful electronic engineering; noise is best reduced by proper circuit design and component selection. People vary in their tolerance to noise and poor frequency response. Some prefer to retain as much as possible of the natural brilliance and tonal quality of the music, even if this means a relatively high hiss level; others can't abide hiss and are prepared to accept a more mellow music tone. So an amplifier (or whatever) that provides a good frequency response, together with good tone controls and filters, should be able to satisfy most people.

Not all equipment is as well designed as this, though, so if you come across a signal to noise ratio that seems surprisingly good, check the frequency response: you may find that it is surprisingly bad.

Distortion

Distortion is usually reserved as a term for the type of defect that adds new, specific unwanted frequencies to the original frequency.

As chapter 2 described, the unique character of a musical instrument is due partly to the number and size of the harmonics of the fundamental frequency that is being played. Distortion changes this harmonic structure by adding new frequencies or by changing the size of existing ones; so it can change the character of the sound. There are two main types of distortion: harmonic and intermodulation.

Harmonic distortion adds harmonically-related frequencies to the original frequency. The ear is rather tolerant of such harmonic distortion; most of western music is built up round harmonically-related intervals. And as most instruments produce quite strong harmonics the addition of a small amount of harmonic distortion is not going to change the relative level much, so is not going to change sound quality much either.

Intermodulation distortion is caused by the interaction of two or more fundamental frequencies – and can be quite complicated.

Intermodulation products are not as directly related to the original signals as is harmonic distortion, so they may be more noticeable. In recent years, reviewers have been measuring intermodulation distortion more and more, as the amounts of harmonic distortion in products has been getting less and less (and more and more irrelevant to sound quality).

To be able to compare distortion measurements, you would need to know what form of distortion was being measured and how; the frequency of the measuring signals and their level; how the input and output terminals are loaded might also affect the result. Even with all this information, it is not really possible to say that the lower the distortion level the better the sound quality; it is notoriously difficult to correlate sound quality with the results of distortion measurements.

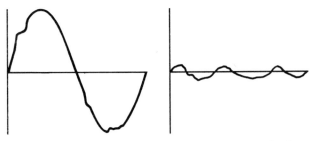

There are two main methods of measuring harmonic distortion. In one, a pure sine wave is passed through the equipment – what comes out will be distorted (*left*). Subtracting the pure sine wave from the output (in the jargon, *notching out the fundamental*) leaves only what the equipment has added. As well as the total harmonic distortion, this **residual**, (shown magnified, *right*) also contains any hum and noise that the equipment has added. A meter reads two values: the total output, which is the fundamental plus noise and distortion and the residual, which is noise and distortion. The distortion level relative to the total output can then be expressed as a dB ratio or, more usually, a percentage ratio.

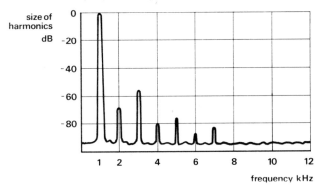

A better method to use is *spectrum analysis*. A machine analyses a waveform and displays the relative level of all the sine-wave frequencies that it is made up from. In this example, the odd-order harmonics predominate, and anything higher than the seventh harmonic is negligibly small. Total harmonic distortion (THD) can be calculated from the display, and will not include any noise or hum.

Spectrum analysis allows the reviewer to examine the relative size of the different distorting harmonics – some experts say that particular harmonics are more objectionable than others – and to measure intermodulation distortion.

Wow and flutter

Variations in the speed with which a turntable rotates a disc, or a tape deck moves a tape, will obviously affect sound quality. In an extreme case, a slowing down from the normal speed will cause the pitch to drop, and voices for example to become slow and slurred; a speeding up will cause the pitch to rise and voices to sound squeaky. Such extreme variations from the true speed should not occur – but even small variations can give noticeable problems.

Wow and flutter are both terms for variations in speed. The only difference between them is that wow is the term used for variations of low frequency (between ½Hz and 10Hz), and flutter is used for higher-frequency variations (between about 10Hz and 100Hz). Note that we are talking about the frequency, or rate, of the *variation* in tone, *not* the frequency of the tone itself; for example, any piano note, whatever its frequency, can exhibit wow.

Because only relatively slow variations in pitch are noticeable (when the variations exceed about 100Hz, they are not audible), wow and flutter are most noticeable on long, sustained notes – particularly from piano or woodwind instruments. Wow is the more directly obvious of the two, and can give a drunken effect to the music, especially in extreme cases. Flutter tends to spoil the subtleties of the music, clouding the sound. In extreme cases it gives a burbling effect to woodwind.

Making the measurement. Despite their audible differences, wow and flutter are generally measured together. The measurement is usually weighted to give more prominence to the frequencies where wow and flutter are audible. The amount of variation often itself varies over time – it is possible either to measure the largest regularly-occurring peaks, or simply to measure the average variation.

The averaging method, usually *weighted root-mean-square (WRMS)*, gives results that are roughly twice as good as those given by a peak measuring method and is the method you will generally find in specifications. But a peak weighting – *Which?* usually uses the DIN peak weighted method – generally correlates better with what people hear. Watch out for this in performance data.

Sensitivity

Sensitivity specifications are often given in hi-fi. In essence, sensitivity is a measure of how much you get out for what you put in – or, alternatively, a measure of how much you have to put in to get a certain amount out.

What you put in or take out – and into and out of what – vary depending on what you want to measure. For example, the sensitivity of a loudspeaker is a measure of how loud a sound you can get out of it for the electrical power (from the amplifier) that you put into it. Alternatively, you could specify how much power you have to put in to get a certain loudness of sound out.

Making the measurement. Methods vary a lot, depending mainly on what piece of equipment is being measured; the individual chapters explain which sensitivity measurements are useful and which are not.

Chapter 4 Loudspeakers

Starting with loudspeakers – which come at the end of the hi-fi chain – is not as daft as it might appear. For a start, it has always been the experience on *Which?* that the loudspeaker is the item that has the greatest bearing on the overall quality of the sound you hear.

It also makes sense to choose your speakers before the other components: this allows you to pick an amplifier of the appropriate power, and a pick-up cartridge with a sound quality that will complement that of the speakers.

Speakers are perhaps the only part of a hi-fi system that look pretty simple. They're not, usually. Nevertheless, a description of **how they work** need consider just two elements in the first place – the chassis (or drive unit) and the cabinet.

Drive unit

This is the main component of a speaker, the part which actually produces the sound. Most speaker drive units work on the same principle. The varying electrical signal from the loudspeaker output of the amplifier is fed through a coil of wire in the drive unit which is suspended in the middle of a strong permanent magnet. As the electrical signal varies, the coil moves within the magnet, following the variations of the signal, so that a large signal will give a large movement, a rapidly-varying signal a rapidly-varying movement. The coil is connected to a light cone, made of paper or plastics, which moves with it pushing the air, and so producing sound. This is the **moving coil cone loudspeaker**.

A few designs work on a different physical principle – **electrostatic**, rather than electromagnetic. If two electrically-conducting sheets of material are placed facing each other with a small insulating gap in the middle, and an electrical voltage connected between them, the sheets will try to move together. If the sheets are large, with one rigid and the other light and flexible, and a varying electrical signal from the loudspeaker output of an amplifier is connected across them, the flexible plate will move backwards and forwards in the same way as the cone of a moving-coil cone loudspeaker, so producing sound. Electrostatic speakers' chief disadvantage is that they require their own mains power supply.

MOVING COIL CONE SPEAKER

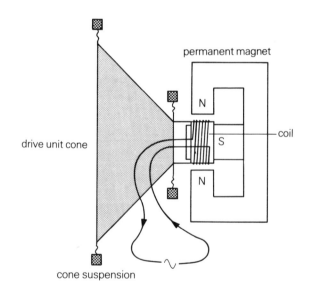

The coil of wire, through which the output of the amplifier passes, is suspended between the poles of a powerful permanent magnet. To it is fixed the drive unit cone – the cone surround and the suspension at the base of the cone hold both the cone and the coil in place. The springiness of these suspensions is an important part of the design.

In practice, drive units are rather more complex than this simple description.

A single drive unit is not usually enough to cover the whole of the audible frequency range. For the lower frequencies, a drive unit has to be large, so that it can move a sufficient amount of air; at higher frequencies, a drive unit has to be small and light so that it can respond quickly to the fast-changing waveform. Most speakers use two drive units – a **woofer** to handle the bass and mid-range frequencies, and a **tweeter** to handle the high

ELECTROSTATIC SPEAKER

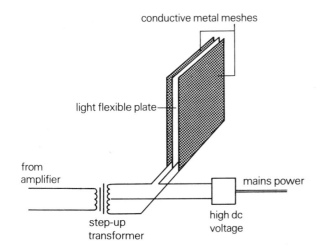

A practical electrostatic loudspeaker has a flexible plate of very light, thin material positioned between two conductive metal meshes. A large, steady electrical voltage (a *polarizing voltage*) is connected to the meshes so that the voltage between the plate and each mesh is the same, which makes the plate stay at an equal distance from each mesh. The audio signal is applied to the meshes: this unbalances the voltages and causes the plate to move towards one or other mesh in sympathy with the audio signal.

A step up transformer at the speaker's input is needed. This increases the audio signal voltage to make it large enough to move the plate, and at the same time reduces the speaker's impedance to make it small enough to be driven by ordinary amplifiers.

frequencies. Some speakers use a separate unit to handle the mid-range (which is sometimes called a squawker – thankfully not too often, since the last thing a loudspeaker should do is sound as though it is squawking). Some split the high frequencies into two, and use separate tweeters and super-tweeters. Be warned: there is no guarantee that the more drive units a speaker has, the better its sound quality – indeed, the more drive units, the more problematical the design process is.

Sub-woofers are extra speakers for very low bass – worth having only if you play music with a lot of bass, and if the rest of your equipment can handle it.

To ensure that each drive unit is fed with only the range of frequencies that it is supposed to handle, electrical circuits are built into the speaker – this is the **cross-over unit**. The frequency response of the complete loudspeaker depends a lot on how well the cross-over unit does its job of reducing the signal to one drive unit, and increasing it to the next, at the cross-over frequency. The better it does the job, the smoother the frequency response is likely to be.

Cabinet design

Drive units on their own do not produce particularly good sound. This is mainly because, as the cone moves backwards and forwards, it produces sound waves from both its front and back. Some of those from the back find their way round to the front – where, because they are *out of phase* with those from the front (see page 20), they tend to cancel each other out. This becomes a real problem at lower frequencies, resulting in a sound which is very weak in bass.

One answer is to find some way of preventing the sounds from the back getting round to the front. The simplest solution is a large board placed in front of the drive unit (with a hole cut in it to let the sound from the front out). Such a *baffle*, though, would have to be impracticably big to be completely effective – over 35ft across to reproduce bass down to 30Hz.

A sealed box also prevents any of the back sound from mixing with that at the front: because such a cabinet would behave like an infinitely-large baffle, it is often known as an **infinite-baffle enclosure** (or *acoustic sus-*

CHAPTER 4

SPEAKER CABINET DESIGNS

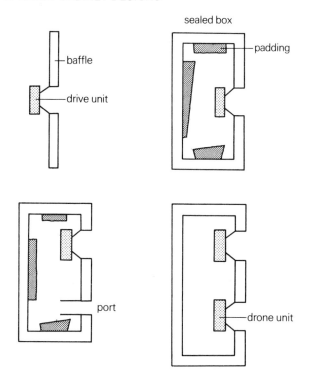

The baffle (*top left*) prevents some of the sound from the back of the drive unit from getting round the front and the two tending to cancel each other out; the bigger the baffle is, the better it is at doing this.

A sealed box (*top right*) prevents any of the back sound from getting round the front, and so acts like an infinitely-large baffle. Note the absorbent padding on the walls to damp out cabinet resonances.

In a bass reflex design (*bottom left*), the back sound is deliberately allowed to get round the front, but in a controlled way: how well controlled depends on how good the design of the port is.

The auxiliary bass radiator (*bottom right*) is a type of bass reflex enclosure in which the port is replaced by a *drone* unit – basically a non-driven loudspeaker cone which moves in sympathy with the bass drive unit, so enhancing bass output.

pension). Its main disadvantage is that, as the drive unit cone moves backwards, it has to expend energy squashing the air in the box, since the air cannot get out. This in turn tends to reduce efficiency – that is, a large number of electrical watts from an amplifier are needed to produce enough acoustical watts for a loud enough sound. This was a problem in the days of the valve amplifier, but with transistors, amplifier watts are relatively cheap, so poor efficiency does not matter very much.

The main advantage of the infinite baffle loudspeaker is that it can be made very small, and yet still have a reasonably good bass response. But there's no escaping the limitations: for a given response, the smaller the cabinet the less efficient the speaker can be.

On the principle that, if you can't beat them, join them, an alternative approach to getting good bass is to arrange for the sound from the back to reinforce, rather than cancel, the sound from the front – this makes the efficiency rather higher than for an infinite baffle. There are various cabinet designs for doing this – all with some sort of hole in the front panel for letting the rear sound out. Unfortunately, it is really not as simple as that sounds: the hole has to be properly sized and positioned, and 'loaded' before reinforcement will take place. So, design is critical.

The most usual type of non-infinite baffle speaker is the **bass reflex** (or *damped reflex*). Here, the sounds from the cabinet are fed out through a tube or 'port' in the front panel: the port is designed so that the low frequencies reinforce. A variation on this is the **auxilary bass radiator**: here the port is replaced by a non-powered speaker-type cone unit which moves the air in the cabinet in sympathy with the bass drive unit at low frequencies. This can produce a similar result to a bass reflex design, but in a smaller cabinet.

Horn loaded speakers – although more efficient – are not used much now that efficiency is no longer at a premium. Some people like the colouration horns produce.

Cabinet design, therefore, is not just a simple job of making a neat wooden box. Ports need particularly careful calculation – but even relatively simple infinite-baffle cabinets must be accurately made to a particular size and matched to particular drive units. It is also important that the insides of the cabinets are correctly covered with materials to help damp out any sound resonances that may be set up in the wooden box.

Alternative designs

Another approach to design is to hand over some of the problems to electronic control. In an **active loudspeaker**, a separate amplifier is used for each drive unit – the theory being that the problems of the cross-over unit will be eliminated. These designs are usually very expensive. A **motional feedback loudspeaker** also has built-in amplifiers – the woofer is linked to the amplifier driving it, in a way that is supposed to help reduce distortion, and achieve a better frequency response.

Omni-directional speakers – drive units pointing in all directions, to replicate natural, reflected sound – have not done well in *Which?* tests. The stereo image is destroyed.

Don't be seduced by claims about **monitor** speakers. They're used by engineers for all sorts of reasons – not all to do with sound quality.

Some speakers claim to have **time-delay compensation** (also called *phase compensation* and other names). The various frequencies of a complex piece of music are reproduced by a speaker at slightly different times from each other compared to the original, and the effect is supposed to be distortion and poor stereo image. However, time-delay compensation doesn't appear to be essential: many speakers without it have a good reputation for distortion and stereo image.

How much does design type matter?

Throughout this book you will come across the statement that no one type of design is necessarily better than any other; this is true of loudspeakers.

Certainly, some of the technical points – for example sensitivity – and some of the features of sound quality may be easier to arrange (or harder to get rid of) in one type of design than in another. But there is no single type of design that offers all the advantages and none of the disadvantages – even in theory (and if there were, it would be very easy in practice to fail to get the solution exactly right).

There are literally hundreds of models of loudspeakers available in the shops these days. Partly, this is because loudspeaker sound can be very personal, and manufacturers like to appeal to all tastes. But it might also be because loudspeakers appear to be fairly easy to make: units can be bought in ready-made, and cabinets can be easily constructed from veneered chipboard. Unfortunately, it is not so easy to design a *good* loudspeaker: many firms now find it essential to use highly-sophisticated computer and other technical aids, so that they can examine in detail how their speaker units and cabinets react to the difficult-to-reproduce transient sounds.

On the other hand, loudspeaker design is still something of an art, rather than a science – and there are indeed speakers on the market whose designs are up to 20 years old, and which can still out-perform the majority of modern designs.

Choosing loudspeakers

Deciding on which loudspeakers to buy is a bit more difficult than choosing the other pieces of hi-fi equipment. Not only are the results of technical tests even less meaningful, but a judgement of sound quality in a listening test is rather more of a personal matter. It isn't altogether laziness on the part of hi-fi writers when they refuse to tell you whether a particular loudspeaker is good or not, and tell you to go and listen to it for yourself – it really is impossible to be dogmatic about it. On the other hand, advice is not quite as difficult to give as some reviewers would have you believe.

The results of carefully-controlled *Which?* tests on loudspeakers show that most expert listeners agree on the *type* of sound that a particular loudspeaker has. And generally, they do agree on *how good* the sound quality is. But occasionally, there are differences between what even experts think is a *nice* or *correct* sound, and it is likely that less expert listeners will disagree more often. And however you try to ensure that the rest of the chain is neutral in sound colour, it is likely that the speakers will sound slightly different when connected to different systems. Finally, the room in which the loudspeaker is placed can affect its sound quality quite crucially – see page 41.

Looking for guidance – from *Which?* and other magazines, dealers etc – is subject to the same drawbacks as described for hi-fi generally in chapter 1. But they may help in identifying a short list to listen to – preferably in your own home.

Bur first, you need to take into account **matching**: basically matching your amplifier, but also your room. The key considerations are *sensitivity*, *power handling*, *maximum sound level* and *impedance*.

Sensitivity

Most modern loudspeakers are very insensitive – only about one per cent or less of the electrical power fed to them by the amplifier is turned into acoustical power. Fortunately, amplifier power is these days relatively cheap, and not much acoustic power is needed to get a loud sound.

Sensitivity is often expressed as the sound output in dB (ie how loud a sound is possible) for an input equivalent to one watt. Even if this basic form is followed, it is still difficult to compare specifications. The output measurement can be weighted (ie more emphasis placed on some frequencies than others) or it may be flat; the input signal may be a sine wave of a single frequency, or noise covering all frequencies (again, the noise may be weighted in some way); and the measurement may be made in an ordinary room, or in a special *anechoic chamber* – a room in which the walls, floor and ceiling absorb all the sound that hits them, and do not reflect any as a normal room does.

Rather more useful than a straight specification of sensitivity would be to know how large an amplifier is needed to get a decent level of sound in an ordinary room. Clearly this is a rather nebulous specification, but many manufacturers do manage to give an indication of the minimum amplifier power they think is needed to drive the speakers; this is often around 15W, but can vary from 5W to 50W. *Which?* reports on loudspeakers always give these figures, and you can find them in many manufacturers' specifications and some reviews. It seems, from *Which?* tests, that it would be sensible to pick an amplifier that has rather more power than the minimum, especially if this minimum is as low as, say, 5 or 10 watts.

Do not be fooled into thinking that sensitivity has much to do with loudness: a high sensitivity speaker isn't necessarily capable of being played very loudly, despite what some ads imply.

Sensitivity specifications are difficult to compare, and a quote of 'minimum amplifier power needed' is more useful. In most cases, it is not necessary to worry too much about minimum power requirements – a good working figure is 20 to 25W for the more sensitive loudspeakers, and around 40W for the less sensitive designs. In general, go for more power than you think you need (but be careful that this isn't so much as to maybe damage your speakers – see the next section).

Power handling

If you drive a speaker too hard, you might damage it – permanently. To warn you of this, all speaker manufacturers quote, in some form or another, the maximum power handling capability of their speakers.

The figure is only a warning. Do not think that it is in any way a mark of the quality of the speaker: a 100W speaker need not sound any better or louder than a 20W speaker when connected to the same amplifier. Indeed, a 100W speaker connected to a 100W amplifier need not sound any louder than a 20W speaker connected to a 20W amplifier (if the 100W speaker is much less sensitive than the 20W one – which will often be the case).

As usual, there are different methods of expressing the maximum power handling capacity of a speaker. Probably the most sensible is an unambiguous declaration of the *maximum amplifier power* rating you should use. Many manufacturers do quote this figure, and so do *Which?* reports. It might be on the cautious side – but if you stick to it and your speakers still suffer damage from overloading, at least you will have a good case for thinking that they, rather than you, were at fault.

Another measure is safe *maximum continuous power*. This usually yields a figure well below what the speaker would be capable of handling in practice. This is because

LOUDSPEAKERS

what damages a speaker is not power itself, but a combination of the level of power and the length of time the speaker has to handle this power – so a speaker can handle higher power than its continuous power rating implies if it's only for short periods. As music does have short peaks of much greater power than the average level, an amplifier's power output rating could safely be more than the speaker's maximum continuous rated power handling capacity.

How much greater, though, is difficult to determine. It depends partly on the type of music. Rock, pop and electronic music usually have an average power that is closer to their maximum power than most classical music, so the leeway you could allow is less. In any case, amplifiers are also rated in terms of continuous power output, and are often capable of delivering more power than their rating during short peaks. All this means that it is not a good idea to expect a speaker rated on continuous power to be able to handle the output of a much larger amplifier. As a good rule of thumb, assume that double the continuous rating is about the most you could get away with in safety.

There are standard tests in which the power handling capability of a speaker is measured using a short burst of power repeated over a long period. These are little used, but if a specification quotes *DIN power handling capacity*, or *music power capacity*, it would be wisest to treat these as the maximum safe amplifier rating.

Of course, using an amplifier of a power higher than a speaker is designed for will not necessarily damage the speaker, unless the volume is turned well up. So if you think you can exercise restraint, a larger amplifier should still be safe. Paradoxically, an amplifier that is too small could also damage a speaker if it is driven beyond its power limits.

All manufacturers give a figure for the largest maximum power handling capacity. This gives no information on how good the speaker is – it's simply a warning to help prevent you from damaging the speaker. Nor on its own can it tell you how loud the speaker sounds – this depends also on sensitivity. Too small an amplifier, if driven hard, is as likely to damage a speaker as too large an amplifier.

Maximum sound level (loudness)

In theory, the greater the sensitivity and power handling capacity, the louder the speaker can be made to sound. In practice, how much power a speaker can take is likely to be determined not by its power handling capacity before damage occurs, but by the amount of power it can take without causing audible distortion. This depends, as usual, on the type of music being played and to some extent on the listener – some music and listeners are more tolerant of distortion than others. The room will also affect how loud the speaker sounds – see page 41.

This all means that any indication of maximum loudness can be only a guide to how loud a speaker sounds – and anyway, most manufacturers do not give a figure. It also means that comparing results across different reviews is difficult: even if they used the same listening room, music, and so on, the listeners' idea of what distortion is tolerable may well vary.

Does it matter very much? Maximum loudness can certainly vary a lot – the loudest speakers can sound four times as loud as the quietest. (To put this another way, if you were content to drive speakers capable of making the loudest sounds at the same sound level as the quietest, you would need only about one twentieth of the amplifier power – if they had the same efficiency.)

On the other hand, in *Which?* listening tests, all speakers seem to be capable of being driven to fairly high levels, even in a biggish listening room. So unless you are after particularly high volumes – for rock music, perhaps – most speakers should sound loud enough. If you do want very loud sounds, though, look for reviews that quote a maximum sound level. Treat the figures as rough, comparative ones, and try to find one review which covers all the models you are interested in. *Hi-fi Choice: loudspeakers* includes figures for maximum sound level.

Although sensitivity and power handling capacity do affect loudness, it is not possible to work out exactly how. But a speaker with a higher maximum power capacity *and* greater sensitivity should be capable of being driven to higher volumes than others. But, for many people, maximum loudness will not be an important consideration.

Impedance

The speaker impedance is its electrical load, as interpreted by the amplifier output. If this load is much lower than the amplifier was designed to drive, then the amplifier could be damaged. A higher level will not cause any damage, but it will restrict the power output of the amplifier, leading to a lower maximum sound level.

All manufacturers quote their speaker impedance – usually 8ohm; sometimes 4ohm (especially for Continental models). A very few are quoted as 6ohm – but it is usually easy to categorise these as 8ohm or 4ohm. So the matching rule is quite simple: pick an amplifier that is recommended as suitable (by the amplifier manufacturer) for driving speakers of an impedance at least as *low* as the impedance quoted by your speaker manufacturer. In practice, all amplifiers will cope with speakers of down to 8ohm impedance, so there is a problem only if you have 4ohm speakers – in which case, the amplifier has to be capable of driving speakers as low as 4ohm. Few amplifiers are capable of driving two pairs of 4ohm speakers at the same time. But as long as you are happy to drive only *one* pair of 4ohm speakers, this should not restrict your choice too much.

As usual, there is a slight added complication. The speaker impedance is only *nominal* and can vary a lot with the frequency of the signal fed to it – see the graph alongside. This is all right, as long as the minimum value is not too far below the nominal. If the manufacturer or reviewer prints an impedance curve, you can check this for yourself – but there is no need to be too pedantic about it, and few speakers that are rated at 8ohm will prove unsuitable for use with 8ohm amplifiers.

If the speakers you select are rated at 8ohm, there is no problem about selecting a suitable amplifier; if they are rated at 4ohm, then you must have an amplifier suitable for 4ohm speakers.

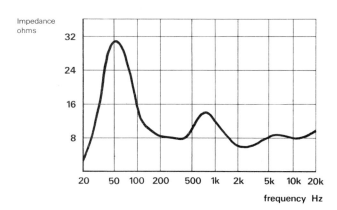

This graph shows how a speaker's impedance varies with frequency: it is often very much higher than the nominal 8ohm value, and sometimes a little below it. The peak at low frequencies is due to a *resonance* in the bass unit – this is a function of the weight of the speaker cone, and the springiness or *compliance* of its suspensions (a similar mechanism is described in chapter 6, for pick-up cartridges). One of the important parts of speaker design is concerned with coping with such resonances – either by using different drive units or by loading it in some way – for example, an infinite baffle enclosure will affect the stiffness of the drive unit's suspensions, and so affect the resonance.

The way in which the impedance changes can be a pointer to how difficult the speaker might be for an amplifier to drive – though *Which?* tests suggest that loudspeakers cause less problems for amplifiers than some people think, especially perhaps if the amplifier is capable of driving loads down to 4ohm.

Finally, note that this is not a graph of frequency response, and tells you nothing about how the speaker will sound.

Listening to speakers

By far the most important factor in a speaker's sound quality is its **colouration**.

Colouration is partly to do with frequency response – and a flat frequency response is important for good sound quality. It is useful if the response is flat even if you are not looking directly at the speaker, but are at an angle to it (in the jargon, the *off-axis response* should be flat).

But even if the response is flat, a speaker may still possess a distinct tonal quality, which can be due to a number of factors, such as electrical resonances in the cross-over unit, and acoustic resonances in the cabinet (and in the listening room). Colouration is usually taken to mean this tonal deviation from the original.

Because it is a deviation from complete naturalness (or *trueness to life*, as the vogue phrase has it), it is usually a bad thing – but some colouration can make a speaker sound mellow and warm, or hard and bright, or exciting; and this sort of thing may be more to your liking than complete naturalness.

Colouration can't be measured on instruments – it is essentially a subjective judgement. This obviously makes it difficult to classify a speaker's quality in a way that will mean the same thing to everyone. However, many of the descriptions used are more or less self-explanatory, and will mean roughly the same thing to most people. Many of the terms – such as *nasal*, *chesty*, and so on – are often used as descriptions of the human voice, and indeed speech can be a particularly revealing test of a loudspeaker's quality.

Even so, different reviewers will have their own pet names for different types of sound (rather as wine buffs have their special jargon) – and often the only way to understand exactly what particular reviewers are talking about is to listen to a number of the speakers reviewed, and translate their terminology into your own.

Of course, listening tests can give us more than just impressions of frequency response and colouration. Distortion is often expressed in descriptive language, too – eg *fizzy*. And listeners can describe how good the stereo effect is – whether all the instruments are clearly defined and their positions unwavering, for instance.

The effects of the room

The type of room the loudspeakers are used in, and the position of the speakers and listener within the room, can all have a marked effect on what the loudspeakers actually sound like. So it is often well worth experimenting. What follows is an ideal guide to some of the changes you might get by altering things; unless you have a separate listening room, of course, you are unlikely to find much scope for change in practice.

The first rule is that rooms with **hard surfaces** – large, uncurtained windows, smooth, plastered walls and ceilings, uncarpeted (and uncork-tiled) floors, and little in the way of soft furnishings – will make a speaker sound brighter, possibly with a hard quality. The multiple echoes or reflections that hard surfaces give will make the sound appear louder, but could confuse the stereo image. Most living rooms, though, will have soft surfaces which will absorb at least some of the sound, especially the high frequencies. If your room has carpets, curtains, and sofa or armchairs, it is likely to have reasonably good acoustics for listening to music.

Putting more **soft surfaces** into a room will cut down the echoes even more – turning a room from a live one into a dead one. Too much of this sort of thing can make the sound appear lifeless. Rooms can be made more dead by covering walls, or parts of them, with curtains, carpet, or cork tiles, and by increasing the amount of soft furnishings. This sort of treatment should not be done haphazardly, though – do it only if you are sure that there is a problem, and that it is due to the room. You will need a very self-critical ear to determine whether the changes you make affect the sound quality – and, more importantly, whether they improve it.

The second rule is that the **position of the speakers** can affect the sound that you hear. Walls, particularly hard plastered ones, act as reflectors. Depending on how close the speaker is to a wall, some of the low frequencies from the reflection will reinforce those directly from the loudspeaker by the time they get to your ears – in other words, placing the speaker close to a wall will boost bass. The same thing happens if a speaker is placed close to a floor – and placing a speaker on the floor and in a corner will give the biggest bass boost of all. Some speakers are specifically designed for mounting close to a wall – some, but not all, bookshelf types, for example. These rely on reflections from the wall to help boost their low bass output. But do not be fooled into thinking that more bass is necessarily better – if the speaker's bass is not only weak but sounds bad too, then boosting it will only make the poor sound quality more obvious. A further problem is that as well as reinforcing some bass frequencies, wall and floor reflections will also cause some frequencies to cancel out – so that mounting against a wall can result in uneven bass reproduction.

Often, a manufacturer's instructions will be a good starting point for positioning. If they recommend mounting on open stands to clear the speaker from the floor, and placing away from the walls and corners, try this position first, and move the speakers closer to the walls only if you think that bass could stand being boosted. Note that away from walls and corners often means about two or three feet away – which would make large inroads on most small living rooms. This may be one good reason for choosing small speakers that can be mounted hard up against a wall – but not all small speakers do perform best like this; some need to be positioned just as far away from walls and floors as larger speakers.

High frequencies are relatively directional, and are easily absorbed in soft furnishings, so it is a good idea to make sure that the tweeter is on a level with your ears when you are sitting in your listening position, and is pointing towards you, and that there are no obstructions between the speaker (again, particularly the tweeter) and you.

Finally, your **listening position** can influence the sound that you hear. The usually-quoted rule is that for best results – particularly of the stereo effect – the two loudspeakers and you should be at the corners of an isosceles triangle, with sides from about 6ft to 10ft in length. It is usually best for the speakers to be angled inwards towards you, so that you are listening *on axis*. Putting the loudspeakers too far apart, or getting too close to them, can break up the stereo image – you hear sound coming from the two speakers, but nothing apparently from between them (often called the hole-in-the-middle effect). Moving the speakers closer together, or listening from farther away, makes the spread of sound (the *sound stage*) appear narrower, which you might think is more realistic.

Although the point of the triangle is supposed to be the best listening position, you can get acceptable (and sometimes even better) results by sitting behind this spot – see diagram alongside. If you move outside this area, the most likely consequence is that you will lose the full benefit of the stereo effect – and as you move closer to one side, the sound will appear more lop-sided. The frequency response is likely to change too: high frequencies are likely to be less good.

There is an added complication to all this, though: you hear sound not only directly from the loudspeakers, but also by way of reflections from all the room walls. In much the same way as reflections from behind the speaker can reinforce or cancel particular low frequencies, so reflections all across the room – from wall to wall, floor to ceiling, and so on – can set up reinforcements and cancellations. The result is that, at some points in the room, one or two particular bass frequencies can be enhanced or reduced in volume – so the apparent frequency response of the speaker can vary, depending on where you are listening. So, to ensure that you are getting the best results, you should not only move the speakers around, but also vary your listening position within the room and relative to the speakers. You should be able to find a position that gives good results: if you find that moving about is not causing noticeable differences, then the chances are that the room is not affecting the frequency response anyway, and you can sit wherever is most convenient.

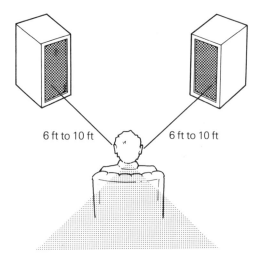

It is generally reckoned that you get the best stereo effect if the speakers are angled inwards slightly, so that they are pointing directly at you, and if the speakers and you form a triangle whose long sides are both 6 to 10ft. The shaded area should also give good results – so more than one person can listen effectively. Note that the lines from the speakers are the *on axis* lines.

TESTS ☑ ☐ ☐ ☐ ☐ ☐

Experts generally agree that many technical tests on loudspeakers are irrelevant, and tell you little about how the loudspeaker sounds. Speaker designers, of course, do need to carry out technical tests – but users may need only listen to them; *Which?* reports on speakers rely entirely on listening tests for judgements of sound quality.

However, you may come across one or two tests in some reviews. A graph of **frequency response** is often shown. This plot is often made by driving the speaker in an *anechoic chamber*, and measuring its output via a microphone. If the test is carried out with the microphone at different angles to the speaker, the reviewer can get some idea of how directional it is – that is, how the response falls off as the angle to the speaker increases. It is rather dangerous to try to predict from a response plot what the speaker will sound like – but it is useful the other way round: it can suggest reasons why a speaker sounds the way that you find it does when listening to it carefully.

A few speakers have **contour controls** to alter the frequency response – usually allowing you to reduce very high frequencies by one or two different fixed amounts. They usually affect a part of the frequency range that amplifier tone controls don't. They may be useful for matching speakers to a room. But the amount of change they offer may be too severe for the sort of tonal correction most people would want to do. So, not a particularly essential feature.

The use of **impedance response** data has already been dealt with on page 40.

Distortion measurements are sometimes shown. The results can look alarmingly high and vary a lot with frequency and power applied; but in listening tests, distortion seems to be less of a worry than colouration and frequency response.

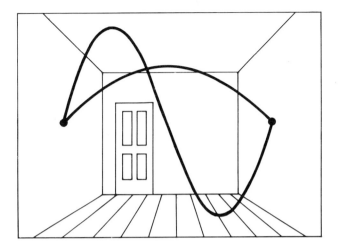

When a sound wave hits a wall in a room, some of it will be reflected – the harder the surface the more the reflection – back towards the opposite wall; when it meets that wall, some of it will be reflected back again, and so on. In any room, there will be frequencies at which the various reflections will have their points of high and low pressure at the same locations in the room – so the wave will tend to build up. Such a wave is called a *standing wave* (because it appears stationary, rather than moving); it gives rise to a resonance effect. The result of such a standing wave is that sound at the frequency of the standing wave will appear particularly loud in some parts of the room, and particularly quiet in other parts.

The first standing wave – in hi-fi and acoustics, often called an *eigentone* – occurs when the frequency of the sound is such that half its wavelength fits exactly into the length of the room. Other standing waves or eigentones will occur at twice half the wavelength, three times half the wavelength and so on. The diagram shows the first two eigentones.

For a room of length about 16ft, the fundamental eigentone will be at about 40Hz, and there will be harmonics at 80Hz, 120Hz, 160Hz and so on. The fundamental resonance is the most important; the others are less audible.

Of course, eigentones will also develop between side walls, floor and ceiling, and between all the various corners. Although the resonances won't go away, it is unusual for them to completely overpower the response unless, perhaps, the room dimensions are such that the various eigentones are at the same, or at harmonically-related, frequencies – hence square and cubic rooms are likely to give the most problems.

Wiring up speakers

Avoid using thin flex for connecting your speakers to the amplifier. Five-amp lighting flex (the sort used to connect lampholders to the ceiling rose) would be a good choice for most short-to-medium runs. For runs over about 20ft long, especially with higher-powered amplifiers, it might be worth using thicker flex – say 13amp. That doesn't mean that this much current will be flowing through the leads (though at very high powers the current could be around 5amps), but it is important for strong, tight bass reproduction that the resistance of the connecting leads is kept low.

Recently, special loudspeaker cables have been sold with glowing promises of how much better they will make your system sound. It is not impossible that they might make a small *difference* to sound quality, but it is unlikely that they can make a system sound *better*. They are usually expensive.

Various connectors are used to connect the cable to loudspeaker and amplifier. The simplest is some sort of screw terminal around which you wrap the bare ends of wires. Banana plugs (sometimes called 4mm plugs) are easier to use if you want to connect and disconnect speakers often. Speakers often have DIN sockets. The male or female plugs are fiddly to solder to wires, so you are likely to buy ready-made loudspeaker cables. DIN plugs and sockets often form quite a high-resistance connection, so they are not too suitable for use at high powers. (The wire in the ready-made cables is often rather thin, too.)

When connecting a pair of stereo speakers, it is important to get them *in phase*. This means ensuring that the cones on the two speakers are either both moving *in* or both moving *out* when the same signal is fed to them. If they are out of phase, so that one is moving in when the other is moving out, the bass frequencies will be weak (because the two speakers aren't pushing the air in unison) and the stereo image will be muddled. To check phasing, switch the amplifier to mono and place the two speakers face to face a few inches apart. Reverse the connections to one of the speakers: that is, at either the loudspeaker or the amplifier end of the lead, swop round the two wires, so that the one connected to the (usually) red-coloured terminal is connected to the (usually) black-coloured terminal, and vice-versa. The connection that gives the louder sound, and the more bass, is the correct one.

You cannot easily try this test if the lead has DIN connectors at both ends, because they are not reversible. On the other hand, they should be wired up correctly, so that the chances of them being out of phase should be small. It is not impossible, though, that the lead, amplifier or loudspeaker has been wired incorrectly, so if you suspect the speakers are out of phase, it is worth trying to check (eg, cut the lead in its centre, and do the connection-reversing there).

It is also not unknown for the different drive units in a speaker to be out of phase with one another – suspect this if one position of the connection gives the better bass, but the other position gives the better stereo image. If the speakers are new, it would be wisest not to try to correct this yourself.

Most amplifiers have switches to enable you to run two or more sets of speakers either individually or together. Usually, the speakers will be connected *in parallel*. Two 8ohm speakers connected this way look like a single 4ohm speaker to the amplifier. Since most amplifiers are capable of driving 4ohm speakers, this is usually no problem. But, two 4ohm speakers connected in parallel look like a 2ohm load – and few amplifiers can drive a load as low in impedance as this.

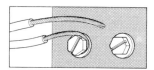

A Screw terminal

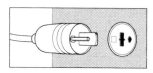

B DIN socket

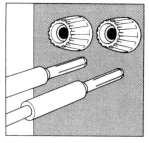

C Banana plugs

LOUDSPEAKERS

This point will usually be covered in the amplifier's instruction book – so ask to see a copy before you buy, if you really want to drive two pairs of 4ohm speakers simultaneously. (When only one pair is switched in, of course, there are no problems.)

If your amplifier has no switches for connecting more than one pair of speakers, the diagrams on this page show that you should find it easy to make your own version – but check first that the amplifier will be able to cope with the lower impedance.

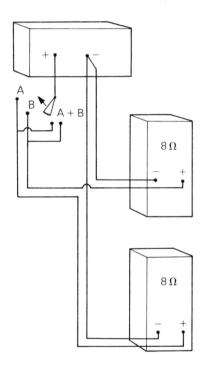

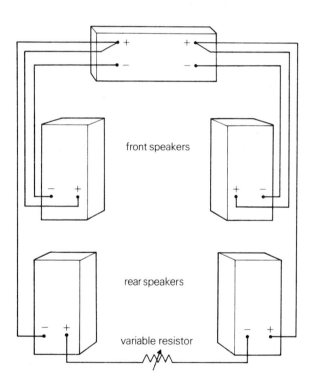

This is how to wire up two loudspeakers so that you can listen to one or the other or both. Only one channel is shown; the other (for the other loudspeaker in each stereo pair) is identically wired. If your amplifier already has a speaker selector switch, it will be wired in this sort of way.

This is one way of connecting a pair of rear speakers to get a form of quadro; how it will sound depends mainly on the method used for recording the music. The resistor (about 500 ohms resistance, and a couple of watts rating) is not essential, but gives control of the rear speakers.

Chapter 5 Amplifiers

In recent years, amplifiers have been the most controversial piece of hi-fi equipment. Some experts claim to hear differences in sound quality between different brands as great as the difference between chalk and cheese. Other experts say that there are no discernible differences among good-quality amplifiers – and back up their opinions with the results of carefully-controlled listening tests. The pro-difference lobby's response to that is that sensitive ears cannot be expected to pick out subtle musical differences in a coldly-clinical listening trial.

Which? tends to belong in the anti-difference camp. Its listening tests are very carefully controlled, as they have to be if the results are to be meaningful, and the numerical part of the data collected is statistically analysed. Yet the tests are conducted in as relaxed an atmosphere as possible, and it is unlikely that the listeners are so put off by the test conditions that they wouldn't notice differences between equipment that they would do in more normal listening. Yet *Which?* rarely finds significant differences in sound quality between different amplifiers – and even when it does, the differences are never huge.

The only way you will find out if you agree with this, of course, is to carry out your own listening trials. If you do agree, you can save quite a bit of money: many of the amplifiers favoured by the pro-difference lobby fall into the expensive (or very expensive) bracket. But be warned: if the results are to be at all meaningful, great care has to be taken – there are details of listening test methods on page 14.

Having disposed, one way or another, of the sound quality of an amplifier, what else is there to consider about it before you can decide which to buy? Probably the most important factor is the *power output* (dealt with opposite) – whether the amplifier will be powerful enough to drive the loudspeakers you connect to it. Then you will need to consider the *features* (opposite) – knobs and switches – that it provides. Details of *input and output sockets* (page 52) are important to ensure that the amplifier *matches* the equipment that is to be connected to it. Most *test results* (page 53), both the pro- and anti-difference lobbies agree, though for different reasons, are of little help in deciding how an amplifier sounds, but they may be important to you.

First, it is helpful to know a little about what an amplifier does.

The amplifier's job

An amplifier has two main jobs to do. First, it takes in low-level signals from record deck, tape deck or tuner. The record deck signal needs special attention: for a start, it is at a lower strength than most of the other signals, so it needs some amplifying; then the frequency response needs correcting, to flatten out the artificial high-frequency boost that all records are made with (see chapter 6 for details of this *RIAA* correction). Most amplifiers have some means for tailoring the frequency response – tone controls or filters that you use to correct for a recording that is slightly dull, or to remove some of the hiss from old records or noisy radio broadcasts. All this signal processing goes on in the *pre amplifier* section.

The signal then passes to the *main or power amplifier*. The job of this section is to provide the brawn – to turn the low-level signal into one of large enough voltage and current to operate a loudspeaker. In a stereo amplifier, of course, both the pre and power amplifiers are duplicated: one for the left channel, one for the right.

Most amplifiers are *integrated* – that is, both the pre and power amplifier sections in one box. Some models, though – particularly higher-powered ones – have them *separated*. This has advantages:

● it allows you to match a particular pre-amp that you like (perhaps because of its facilities) with a power amplifier of the right sort of power rating for your speakers. There is little scope for this within one manufacturer's range, though, because few make a series of different pre and power amplifiers in the way that they do integrated amplifiers

● it allows you to mix different brands of pre and power amplifier. You might want to do this if you believe that amplifiers (pre or power) do differ significantly in sound quality, and you're after the ultimate in sound perfection. Or you may be after a particular set of pre amplifier facilities, but do not want to use that manufacturer's power amplifier. (However, pre amplifiers are often pretty austere affairs, and the problem might be finding *any* pre amplifier with the facilities you want.)

● since the power amplifier contains no knobs and switches (except, perhaps, for an on/off switch), it can be tucked away out of sight and only the pre amp, which is generally a rather smaller and neater component, need be left on view. Power amplifiers can run rather hot, though, so care is needed to ventilate them

● manufacturers may make separates because they feel that's the best technical design for their equipment.

For most people, though, these considerations will not be very important and, unless separates are the only ones to offer you what you want, it will be best to consider integrated amplifiers. Generally, they are cheaper, with more facilities and greater output power for the money.

Power output

The power output of an amplifier is a guide to how loud it will sound. Too little power, and you might find that music is not loud enough until the sound is distorting; too much power could just waste money – used carelessly, it might damage loudspeakers. It is only a rough guide, though. The amount of power needed for a given loudness of sound depends on many factors – in particular, on the sensitivity of the loudspeakers used (see page 38), the size of the room and the type of furnishings in it. For most people, between 15W (watts) and 40W per channel would probably be about right. More power would be needed if you like your music very loud, have particularly insensitive loudspeakers, or have a very large room with lots of soft furnishings. More power does not mean the odd watt or two extra, but an increase of two to four times – maybe to 50W or 100W. Small increases in power are not noticeable.

There is more information about power output in the section on tests, starting on page 53.

FEATURES ○|○|○|○|○|

Most amplifiers sport a large number of knobs and switches, flashing lights and moving dials – and the higher the power output the more features amplifiers tend to have. Few of these features are absolutely essential – though manufacturers seem to need to provide them to help them sell their wares. Indeed, some cult amplifier manufacturers claim that the more switches and signal-processing circuitry there are, the poorer the amplifier will sound; they offer very austere amplifiers with nothing more than an on/off switch and a volume control.

Even if unmoved by the flight-deck approach to features, you will probably want more than the cult amplifiers provide. The important thing to do is to decide carefully which features are *necessary*, which are merely desirable, and which you are completely indifferent to. Use the detailed descriptions below to help you make up your list.

Armed with the list you can then go shopping – either foot-slogging through the hi-fi shops, or in comfort through the manufacturers' catalogues – rejecting those amplifiers which do not have features you think are necessary, and short-listing those that do for further consideration. Since most amplifiers offer more than the basic features, this task should not be too arduous, and you will probably end up with a rather long short-list. An alternative approach is to use your features list as a final check: having found an amplifier which satisfies all your other criteria, check that it has the features you want.

CHAPTER 5

Controls for volume

Most amplifiers have a single volume control, which increases or decreases the amount of sound from both loudspeakers by the same amount. They also have a **balance control**, which changes the relative levels of the left and right channels (that is, gives more, or less, sound from the right than from the left). The effect of this is to swing the apparent source of the sound nearer to one speaker and away from the other. Small adjustments of a balance control away from its centre position may be necessary if, for example, you cannot sit exactly centrally between the two loudspeakers. It is a control that in most cases requires altering only rarely.

Some sets have **separate left and right volume controls** (in which case there is no need for a balance control, as you can alter the left and right levels independently). The two controls may stack one above the other, on the same spindle (*concentric*). These should be *ganged* so that as you turn one, the other turns with it unless you positively prevent it from doing so. In this way, it is easy to change the volume without upsetting the balance. Controls that are not ganged will have to be turned together (and generally by the same *percentage* rather than the same *amount*) if the balance is to remain the same – this is clearly a little more fiddly, particularly if the two controls are knobs rather than sliders (which you can generally adjust one-handed).

Since the volume control is used frequently, it makes sense to look for an amplifier that has an easy-to-identify (eg large) control that is easy to use (either combined left and right, or ganged).

Some volume controls have **click-stops**, so that instead of being able to alter the volume continuously, it can be altered only in small, fixed steps. The steps are usually small enough so that the differences in volume do not appear great. Like many features on domestic equipment, click-stops have been pinched from professional equipment. Professionals use them because it is often important for them to be able to alter the volume by known amounts, and because continuously-variable controls are less reliable than switches. Neither of these considerations is particularly important in the domestic field and in any case, the type of click-stop used on most domestic equipment is not likely to give you these professional advantages.

On the other hand, a click-stop at the mid-point of the balance control – **centre indent** – is quite handy if it has been properly adjusted.

Loudness controls are still popular, despite the purists' objections to them. These are controls which boost the level of the bass frequencies (and usually also the high frequencies, though by a smaller amount) at low listening levels. This is supposed to compensate for the way in which the ear's sensitivity to low and high frequencies falls off as sounds get quieter. The main objection is that it is impossible to design a loudness control that works effectively, since it is impossible to determine how much quieter than 'normal' any particular setting of the volume control represents. In practice, most manufacturers do not bother to try to design an effective control, but simply one that gives instantly-recognisable boosts when switched in. Another objection to loudness controls is that they simply do not reproduce what would be heard live. If you listen to recorded music at a low volume, you want it to sound like live music played at low volume (or heard from afar) – low and high frequency fall-offs and all.

Most loudness controls are simple on/off switches, but some are variable – though most, whether controlled by a switch or knob, are variable in the sense that, as you turn the volume control up, the effect of the circuit should get smaller and smaller, until at normal listening levels there is no boost at all. (In practice, many circuits give large amounts of boost even at high listening levels.)

Fortunately, there are now very few amplifiers with loudness that cannot be switched off – which will satisfy any purist ideals you may have. On the other hand, if you want a loudness control, it is probably not worth searching around for a 'good' one, for the reasons given above: if the amounts of boost at your listening level are not to your liking, then modify them with the tone controls.

A very few sets have a separate **pre amp volume control**. Ordinary volume controls (though located in the pre amplifier section) actually work on the input to the power amplifier. This means that, if the pre amplifier is overloaded with too large a signal, turning down the main volume control cannot reduce the distortion that the overloading is causing in the pre amp. On the other hand, if the pre amp is working with a very weak signal, the main volume control will have to be turned well up, which generally increases noise. A pre amplifier volume control can ensure that the signal presented to the power amplifier is as large as possible without being distorted.

In practice, though, most present-day equipment will not present an amplifier with unduly large or unduly small signals, in which case a pre amp volume control is unnecessary. A few amplifiers have **variable input sensitivity pre-sets**, which allow each input socket to be matched to the signal level of the input that is feeding it. These are more versatile than pre amp volume controls and, as they are hidden away round the back, can be forgotten about once set.

Unless you change equipment frequently, it is not important to look for an amplifier with a pre amp volume control – better to get one that is properly matched to your equipment, or that has variable input sensitivities.

An **amplifier mute** is a switch that reduces the volume – usually by about 20dB. It could be useful if you are diving to answer the phone, for example: when the call is over, you can restore your original listening level quite easily. If the listening level is very high, a reduction of 20dB might not be sufficient for peaceful conversation. Switching in an amplifier mute also gives you a greater range of the volume control when the amplifier is driving very sensitive loudspeakers or if you want to listen at low volumes. So an amplifier mute can be useful, though it is not an essential control.

Volume indicators. There are three main types of indicator that deal with volume, varying in their usefulness. **Power meters** show how much power the amplifier is delivering into an 8ohm load. They are not particularly useful: amplifiers are rarely driving simple 8ohm loads, since speaker impedances change with frequency (and between speakers). Also, the power output of an amplifier is little guide to how loud a sound you will be getting; and in any case, there is no reason to have a meter to *show* you how loud the sound is when you can *hear* it perfectly well for yourself.

Rather more useful are **overload indicators**. These lights (which should react to peak outputs) flash when the amplifier is overloading and thus distorting badly. You will probably be able to hear this yourself, but if you hear distortion when the lights are *not* flashing, then you know some component other than the amplifier is to blame (providing the overload indicators have been properly adjusted).

Amplifiers have protection circuits to prevent the power transistors and the loudspeakers from being damaged if the amplifier is heavily overloaded. (These circuits also disconnect the loudspeakers on switch-on and switch-off to prevent voltage surges through the speakers.) It is quite useful to have **protection circuit indicators** which show when these circuits have disconnected the loudspeakers – then you do not need to worry needlessly that your amplifier has blown up.

Controls for tone

Nearly all amplifiers have two tone controls. One alters the relative level of the *treble* frequencies, to correct for recordings that sound too bright or dull; the other alters the relative level of the *bass* frequencies, to correct for recordings that sound boomy, perhaps. Tone controls, like balance controls, usually have a centre indent. This centre position is called the *flat* (no tone changes) position.

Tone controls are useful, since recordings do vary in their frequency balance. They *might* also help to correct for frequency problems caused by the acoustics of your room. But violent shifts of the tone controls away from the flat positions should not be necessary, and you will probably find that controls are more effective at taming slight excesses of treble or bass than they are at boosting losses.

There are more details about the action of tone controls on pages 55 and 56.

Separate left and right tone controls are marginally useful, because they allow some extra control over room problems. Since they should not need altering very often, the difficulty in operating a non-ganged type is not as important as it is with volume controls. Still, they do introduce a complexity that most people can do without, so don't go out of your way to look for sets with separated tone controls.

Middle frequency controls boost or cut frequencies around 2kHz, which affect particularly the reproduction of speech. There is less need to tailor the middle frequencies than the bass and treble; in any case it is unlikely that any one control would affect precisely the relevant middle frequencies. Again, not really worth going out of your way for.

Graphic equalisers are again a spin-off from professional equipment. Each equaliser boosts or reduces (cuts) a narrow band of frequencies; a number of them together cover the whole frequency range. To be better than just a gimmick, there have to be at least five controls, and each one must affect only the range of frequencies it is supposed to be affecting.

Given all this, graphic equalisers *can* be more useful than simple tone controls, and they will also act as filters (see next column).

A more complicated device is the **parametric equaliser**. You set the frequency band that needs modifying with one control, and the amount of cut or boost with another control. That allows you to select exactly the frequency range to be corrected – but you cannot alter more than a couple of frequency bands (depending on how many parametric controls there are) at any one time.

Tone defeat switches allow the signal to by-pass the action of the tone control section – so that you can compare instantly the flat and tailored responses. Some people also claim to be able to tell the difference in sound quality between the tone defeat position and the tone on position, even with the tone controls set for a flat response – but perfectionists like these are more likely to buy an amplifier with no tone controls at all.

Some tone controls have **switchable turnover frequencies**. The turnover frequency is the point at which the control starts to have an effect on the frequency response. A choice of turnovers can make tone controls more effective – providing the choice you are given is a sensible one.

Click-stops on tone controls are rather more useful than they are on volume controls. For example, they allow you to re-set the tone controls to exactly the same positions you have used before, each time you play a particular record.

Filters are a specialised sort of tone control. They only cut (not boost); they do so steeply; and they do so at the extremes of the frequency range. At least that is the theory – in practice, filters often cut no more steeply than tone controls, and start operating too early (ie at too low a frequency if they are treble controls, or at too high a frequency if they are bass controls).

Some filters have switchable turnover frequencies, like some tone controls. Some also have switchable slopes. Steep slopes have a more violent action than shallow ones. For example, a steep treble filter will reduce the level of high frequencies more than a shallow filter starting from the same point. (More details on slopes and turnovers in *Tone controls and filters*, on page 56.)

Good **filters are useful: they allow you to reduce hiss without removing too much of the brightness and sparkle of the response, and reduce rumble or other sundry bass noises without losing body.**

Controls for stereo

Most amplifiers have some sort of *mode* switch. At its simplest, this switches the amplifier between stereo and mono. Some forms of record rumble, and of hiss from records or radio broadcasts, can be reduced by switching the amp to mono mode, so this is useful. Other functions available on some mode switches are rather less useful. These can include left or right channels only, either through the left or right speaker only or through both speakers; stereo-reverse – left channel through the right loudspeaker and vice versa; and variable separation – a knob which progressively changes the stereo separation from full stereo to full mono.

Quadrophonic, or four-channel functions, are rarely found these days. Some of the cheaper amplifiers might sport a *Hafler-type* quadrophonic switch (under a variety of names), but it is not a feature that is important to look out for.

Selecting sources

Amplifiers have a basic *source selector* which enables you to pick which input is to be amplified and passed to the loudspeakers. It usually selects between tuner, disc and auxiliary (aux) inputs, with tape selection handled by a separate source/monitor switch. (This book usually uses the term *disc* for anything relating to the round, flat, black things played on record decks. *Record* can be confused with tape recording. And *phono* – which manufacturers often use, for example to label disc input sockets – can be confused with the type of plug and socket called phono.)

Disc inputs and selectors are getting more complicated these days, with some amplifiers having positions for up to two *moving-magnet* cartridges and a built-in *head amplifier* for a moving-coil cartridge. Most cartridges work best when properly loaded (see page 69 for details), and some amplifiers now have switchable settings built in.

Complicated disc selectors can be useful; built-in head amplifiers, in particular, can save a lot of money. But, if you actually need these functions, you may find it better to provide exactly what you need with add-on units, rather than restrict your choice of amplifiers.

Controls for tape recorders

Most amplifiers have connections for a tape recorder (cassette, or reel-to-reel, it usually doesn't matter which). These work for recording (off radio, disc or microphone), and for playing back recordings.

Tape controls are often confusing. The *tape out* socket (the one used for *recording*) is usually permanently live; the amplified signal (the one selected by the source switch) is always present, and ready to be recorded at this socket. The *tape in* socket (used for *replay*) does not usually go to the source switch, but to a separate source/monitor switch. With the switch in the *monitor* position, the power amplifier is disconnected from the pre amp section, and connected directly to the tape in socket, for playing tapes. For all other sources, however, the switch has to be in the *source* position. (A tip: if your amplifier appears not to be working, check the position of the tape monitor switch before doing anything else.)

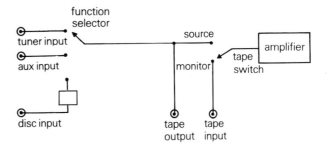

How a standard tape source/monitor switch is wired – separately from the ordinary selector switch.

The reason for this curiosity is that, with three-head tape decks (decks that have separate record and playback heads), you can use this switch to compare easily what you are recording with the recording itself as you are making it. This is called monitoring, hence the name of the switch.

It should be possible to monitor on the few amplifiers where the tape switch is part of the function selector (which is a more logical arrangement), but the changeover from source to monitor may not be so slick.

A few amplifiers have a *record out selector*. With this, you can record from one source whilst listening to another – a boon, and standard on video recorders.

Many amplifiers have facilities for recording and playing back with a second tape deck – though be warned: some sockets marked *tape* are for replay only. If you use an extra deck for replay only, use the *aux* socket.

Dubbing is simply recording on to one tape deck from another. Special *dubbing switches*, or even special sockets for two tape recorders, are not really necessary (you may be able to connect the one you are recording from to the aux input, or you can simply connect the two recorders together directly). But they do save a lot of fiddling about, plugging and unplugging leads. Being able to dub in *either* direction is particularly convenient.

If you want to connect and disconnect a tape deck frequently, you might find it useful to have tape sockets on the *front* of the amplifier.

Complicated facilities for tape recording may be useful to you. But, as with disc selectors, if you have particular needs, it might be more sensible to investigate the range of add-on switch boxes that is available.

Controls (and sockets) for loudspeakers

Most amplifiers have a **selector** for either of two pairs of stereo loudspeakers – one pair in the lounge, and another pair in the kitchen, say. Usually these are called pair A and pair B; some amplifiers can take a third pair, C. The selector can often switch to the various combinations of these pairs, as well as all off.

Many amplifiers will work happily when used with loudspeakers of down to 4ohm impedance (see page 40 for details of loudspeaker impedance), others only with speakers down to 8ohm. Even amplifiers that will take 4ohm speakers probably will not cope with more than one such pair switched on at one time.

Loudspeaker **sockets** are usually either some sort of screw terminals or spring-loaded push-button types into which the bare ends of the loudspeaker cables can simply be pushed – not a great advantage unless you spend a lot of time trying out different pairs of loudspeakers, and they do prevent you making a good, tight connection. Some sets have *DIN* loudspeaker sockets – really too flimsy to be of much use on amplifiers rated at more than, say, 25W to 40W. One or two amplifiers have *phono* sockets: the danger here is that you might accidentally plug in, say, the output from your record deck. Trying to feed the output of an amplifier into a delicate pick-up cartridge is likely to do it a real bit of no good.

Plugging in **headphones** may or may not automatically cut out the speakers. Either way, this could be inconvenient sometimes. Marginally the best solution (because it keeps all your options open) is for plugging in the phones not to cut out the speakers and for the speaker function selector to have an off position.

Other inputs and outputs

Most **microphone inputs** are of the stereo-jack type, located on the front panel of the amplifier. They allow you to use your amplifier as part of a public address system, or disco perhaps. You may be interested in inputs with *variable gain controls*: some of these simply turn the level of the microphone up and down; others will fade out whatever other source is playing (a disc, perhaps) as the microphone is faded up.

Pre-amp sockets let you separate the main and pre amplifiers. There is a variety of possible uses for this, though none is very common. For example, trying out different pre and main amps, for connecting quadrophonic decoders or graphic equalisers, for connecting loudspeakers with their own built-in amplifiers, or for use as tape-recording sockets, to allow you to record programmes after they have been modified by the tone controls.

Plugs and sockets

There are many different types of plugs and sockets used for connecting hi-fi equipment together. The two main types are the **DIN**, used mainly on European equipment, and the **phono**, used mainly on Japanese equipment. Phono sockets are sometimes called Cinch (a trade name), RCA or jack – though almost any plug can be called a jack. Another confusion is that the record deck input on an amplifier is often labelled *phono* – whether the socket is phono or DIN type. (Phono is short for the American word for record deck – phonograph.)

The two main types of socket look entirely different – see below – so if the lead from your tape deck, say, has a different type of plug from the socket on your amplifier, you will not be able to connect the two directly together.

It is fairly easy to buy adaptor leads from hi-fi shops to get over the physical problem. But another problem, which is just as important, is that DIN and phono sockets have entirely different electrical characteristics. Inter-

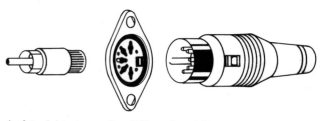

Left to right, phono plug, DIN socket, DIN plug

connecting phono and DIN is therefore likely to lead to trouble. A DIN output connected to a phono input will probably not yield enough input voltage – and so you won't get much volume. Connecting phono outputs to DIN inputs is worse: you are likely to overload the input with too high a voltage, and so get distortion. (You may be able to get round this problem by buying *attenuators* which reduce the voltage at the input and so prevent overloading.)

So if you have to choose between mixing types, or using DIN throughout, it is likely to be better to use DIN. However, using DIN, even exclusively, can lead to problems – if the voltage levels and impedances are badly chosen, the signal to noise ratio could be poor, and some high frequencies could be lost. So it is much better to use phono sockets throughout, if you can.

One crumb of comfort: these problems do not arise with record decks, as the electrical characteristics of their outputs are governed only by the pick-up cartridge, so you can interconnect DIN and phono with impunity here. And one final frustration: occasionally, a piece of equipment (often British) will use DIN sockets, but wired so that they behave like phono ones. So when buying British equipment with DIN sockets, ask the dealer about interconnections.

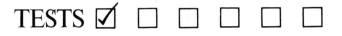

Power

There is a tendency for power to be equated with quality, but a powerful amplifier is by no means better than a less powerful one – it is simply capable of louder sounds. Of course, big is usually expensive, and amplifier manufacturers are more willing to put the latest technology and jazziest features into their more expensive products than into the budget end of the range.

Amplifier outputs are usually quoted in terms of watts of output power. All other things being equal, an amplifier rated at double the electrical output power of another one will be capable of producing double the sound power. As explained in chapter 2, the ear is not sensitive to changes in sound power of less than double (ie increases smaller than 3dB), and to get an increase equivalent to a doubling in loudness level, the sound power has to increase by tenfold (or 10dB). So, if you had a 15W amplifier and wanted something noticeably louder, you would need to go for at least 30W; if you wanted one that sounded twice as loud, you would need to go for 150W (which would probably blow up your speakers – and make your neighbours explode, too).

This all makes rather a nonsense of the practice of most manufacturers, who insist on having a range of very similar amplifiers with only small increases in output power between them.

The use of *watts* in manufacturers' power ratings can cover a multitude of sins – and in the past (because lots of watts looks rather impressive), manufacturers have been clever at finding ways of apparently upping the number of watts their equipment produces, without actually having to increase the output power. The practice is rather dying out nowadays, but it is still as well to be on your guard against it.

The first thing to watch out for is what *impedance* loudspeaker the power output is quoted for. Usually it's 8ohms; but a 4ohm speaker will give a power output figure about 25 per cent greater. Continental amplifiers in particular are often specified into 4ohms; this does not imply that they are bigger cheats, but reflects the fact that many Continental loudspeakers have impedances that are closer to 4ohms than to 8ohms.

The second problem is that the output power available is dependent to some extent on both the *frequency* and the *distortion level* that the measurement is made at. In fact, with modern transistor amplifiers, harmonic distortion stays fairly constant at all power levels until the *clipping point* (see diagram page 54) is reached. And, although the power does drop off at low and high frequencies (see *Power bandwidth* below), this is not very important because full power is rarely demanded at the frequency extremes.

The third snag is the *different ways* there are of *specifying* the resultant power. The most usual way is *watts rms* (strictly, watts average). It is also possible to specify output power in *watts peak* – giving a result *twice* that of watts rms.

This does not exhaust the possible ways of increasing the apparent output. For example, many amplifiers are capable of giving a little more output if only one channel is driven, since the amplifier's power supply has an easier job. So, quoting that figure (rather than that with *both channels driven*) is a little misleading. A sneakier method is to add together the outputs of the two channels – so a *25W per channel* amplifier magically becomes a *50W total* amplifier.

One expression that has more or less died out is *music power*. By using a test signal more like real music than the normal sine wave, it attempted to give a result that was more meaningful to real ears – which is always to be applauded. An irregular signal, rather than a constant one, was used – and because the amplifier had a little time to recover between the bursts of tone, the output was often a little higher – perhaps by as much as 20 per cent over a similar constant-signal test.

As it was a very vague term, its passing is welcome; though recently, interest in this sort of measurement has been revived. Some experts have found that some amplifiers can be driven to subjectively louder volumes than others having the same constant power output. The difference can be that the louder one gives more output on a tone burst test than the other.

In most amplifier specifications these days, a watt is a watt is a watt: most of the differences in measuring technique (such as the distortion level used, whether both or one channel is driven, and whether the result is the figure at 1kHz, or the lowest figure over a range like 20Hz to 20kHz) will not yield vastly different figures for power output. But a few specifications might use music power watts or peak watts or quote a result into 4 ohms – any of which will give a result between about 20 per cent and 100 per cent higher than usual. Watch these don't trip you up when you're comparing specifications.

Power bandwidth

The power bandwidth measurement gives an idea of how severe the fall-off in power capability is at low and high frequencies. The usual method of measurement is to select a level of distortion – say 1 per cent – and plot the power output, at this distortion level, from 20Hz to 20kHz. Since it is not easy to tell the exact frequency at which the power begins to fall off, it is usual to quote the *half-power bandwidth* points – the frequencies at which the power has fallen to half (ie minus 3dB) of its value at 1kHz.

This is a measurement to be taken with a pinch of salt. The results can be affected by the level of distortion chosen – the lower the level, the narrower the half-power range. One per cent is a reasonable level – anything less may give a misleading impression of the amplifier's capabilities. In any case, large amounts of power are not necessary at the extremes of the frequency range, so unless the half-power range is extremely restricted, it is unlikely to be noticeable.

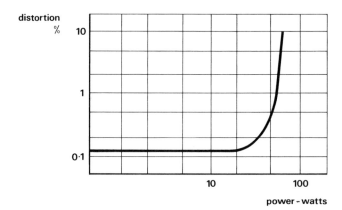

As the output power of the amplifier increases, the distortion also rises – nearly always by very little at first, but later very steeply. With many transistor amplifiers, the 1 per cent distortion is just below the point at which the distortion rises very steeply with little increase in power output – the *clipping point*. Valve amplifiers tend to have a much less steep rise – in the jargon, they go into clipping softly. The hard clipping of transistor amplifiers is supposed to be one of the reasons for the difference in sound quality some people detect between them and valve amplifiers. (But of course, as long as both amplifiers were run well within their power rating, this difference should not be apparent.)

Distortion

An enormous amount of work has been done trying to identify the different sorts of distortion which some people say they can hear in different amplifiers. But no one would be foolhardy enough to claim that distortion specifications – however complex – could tell you how good a hi-fi amplifier sounded (even assuming that you thought that you could tell one amplifier from another).

Don't expect amplifier distortion specifications to tell you very much. The next section describes briefly what the different tests mean.

Types of distortion

There are four main types of amplifier distortion.

Harmonic distortion is still the most usual test to find in manufacturers' specifications, and in many equipment reviews in hi-fi magazines – mainly because it is easy and cheap to carry out. If the type of distortion is not specified, assume it is harmonic.

Intermodulation distortion tests are becoming more popular among manufacturers and reviewers. Unfortunately, there are many different methods, all yielding different results. The CCIF (International Telephonic Consultative Committee) uses two high-frequency signals of the same amplitude – one of the advantages of an intermodulation test. The results of a harmonic distortion test at high frequency are more difficult to interpret, since the distortion products are well above the audible frequencies.

The SMPTE (Society of Motion Picture and Television Engineers) test is less useful for hi-fi equipment.

A more complicated test (often called *swept frequency test*) is much more searching, but the amount of distortion can vary depending on the distance between the low and very high tones used.

Transient intermodulation distortion has been blamed over the last few years for amplifiers sounding nasty. Tests of amplifiers' ability to cope with loud, sudden signals – cymbal clashes, typically – without distorting are difficult to do, and the methods are various. In practice, you are unlikely to hear such distortion with even relatively poor amplifiers.

Transient harmonic distortion. In the past, *Which?* has used a simplified transient test, passing a sharp transient through the amplifier, and measuring the level of harmonic distortion over a period of a few seconds after the pulse has passed through. The distortion could reach relatively high levels (perhaps 0.4 per cent, sometimes very much higher) and last for some time after the pulse had stopped (as much as 0.1 sec). *But* it requires a pulse driving the amplifier nearly to full power to produce these apparently poor results; an amplifier with plenty of power to spare will behave very much better.

Frequency response

A frequency response graph looks rather like a power bandwidth one, but the two should not be confused. For a frequency response measurement, a low power level is used, and the distortion does not have to remain constant. The purpose of the test is to see whether the frequency response remains flat at normal listening levels.

Except perhaps for the disc input, most amplifiers have extremely flat responses from very low frequencies up to very high ones – perhaps 80kHz and beyond. Since most people's hearing stops well short of 20kHz, you might ask what the point of such an extended upper-frequency limit is. The answer is, very little. Most experts agree that it can be useful for the response to extend as far as, say, 40kHz. If the response cut off were lower, the amplifier would not be able to reproduce very fast *transients* (the very sharp beginnings of musical sounds) accurately. On the other hand, if the response is much wider, there is more chance of audible intermodulation distortion frequencies being generated. (There is a saying that hi-fi writers are fond of quoting: the wider you open the window, the more muck can fly in.)

The frequency response from the disc input is not always as ruler-flat, for two main reasons. First, there has to be a correction for the *RIAA recording characteristic* (all records are made with an artificial high frequency boost and bass cut – see page 62). It may not be precisely the reverse of the RIAA recording curve. The result will be a response that is not flat. Second, the voltage that a pick-up cartridge generates across the disc input can depend on the impedance of the input; this varies with frequency, so the response can vary with frequency, too.

If you think you hear differences between amplifiers on their disc inputs, this is the first thing to look into: often the differences will disappear if the amplifier's input impedance is changed to suit the cartridge better.

In rigorous, side-by-side listening tests, variations in frequency response of as little as ½dB might be noticeable. In more relaxed listening, a variation of 1½dB might be tolerable.

Tone controls and filters

Most people will find tone controls useful on occasions, but they do not need to give a particularly wide variation in response. When manufacturers give specifications for their tone controls, it is usually as a graph – perhaps because a set of tone control curves looks rather impressive. Though the curves can give some useful information, you really need check only that the maximum cut and boost is likely to be sufficient – say 10dB each way. Instead of a graph, the cut and boost at particular frequencies may be quoted – around 60Hz for the bass control, 10kHz for treble, would be useful frequencies.

High filters should reduce hiss and disc surface noise, and *low filters* should reduce rumble, ideally without affecting the quality of the music more than is necessary. The usual problem is that filters do not cut steeply enough. Slopes of 6dB per octave (see right-hand diagram opposite) often only duplicate the action of tone controls; slopes of 12dB, or perhaps 18dB, would be more effective in cutting rumble or hiss. Even if the filters cut steeply enough, often the *turnover frequencies* (the frequencies at which the filters begin to take effect) are badly chosen. Look for bass filters with a turnover frequency of less than about 80Hz, and treble filters with a turnover of 6kHz to 8kHz. If more than one of each filter is fitted, it is helpful if the turnover of the second low filter is around 40Hz, the second high, about 9kHz.

On the other hand, filters with slopes that are too steep can upset an amplifier's performance, giving some distortion effects. This depends on the design of the filter circuit, as well as its slope.

Signal to noise ratio

Measuring, expressing and comparing amplifiers' signal to noise ratios illustrate all the problems discussed in chapter 3. The latest American IHF (Institute of High-Fidelity) Standard for amplifiers is very logical. Inputs are loaded realistically, with typical input levels, the volume control set to give a typical output level. Under these conditions, with CCIR ref 1kHz noise weighting, a ratio of 70dB or more for the disc input would probably be as good as needed (comfortably above the 67dB or less of most discs). Many people would find that even 64dB or so would be quiet enough.

If the measurements are made with the volume control turned right down, the rating is effectively that of the power amplifier alone, and is commonly called the *residual* hum and noise. This unrealistic result usually gives a better-looking figure than one which includes the noise from the pre amplifier section as well – so manufacturers may prefer to use it in their specifications. The result may even be based on figures obtained with nothing connected to the input.

If these figures are then related to the amplifier's full *output*, a large, impressive-looking signal to noise ratio can result. Similarly, manufacturers who use full output when quoting the signal to noise ratio of the whole of their amplifier will produce better-looking results than those following the IHF Standard. A 30W amplifier could gain another 3dB, a 100W amplifier, 20dB. Using a less-strict noise weighting curve can also bump up the result – for example, using A weighting might add on an extra 5dB.

With specifications that do not give the measurement method, assume that something like this has been done. So, for a 100W amp, a figure of 70dB would become the manufacturer's 95dB.

Signal to noise ratio measurements on amplifiers are particularly difficult to compare. But it is fairly rare for amplifiers to be the guilty party if noise is noticeable in a system – particularly on inputs other than disc.

THE FAMILY OF TONE CONTROL CURVES

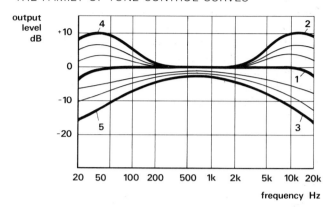

HOW FILTERS WORK

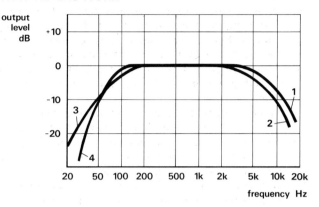

Each of these curves shows the frequency response of an amplifier (from 20Hz to 20kHz) for different settings of the tone controls, giving a *family of curves*. It is usual to show on each curve the effect of boosting (or cutting) both treble and bass controls together, though in practice, of course, you may alter only one of them at a time.

Curve 1 shows the response with the controls in their mid-position – should be a flat response. Curves 2 and 3 show what happens to the response when the treble control is wound fully up or down: the frequency response rises or falls to make the sound brighter or duller. At 10kHz, the response is cut or boosted by about 10dB, which is perfectly adequate. Curves 4 and 5 similarly show full boost and full cut on the bass tone control – again, about 10dB at 60Hz.

Note that the bass and treble boosts start falling at about 30Hz and 18kHz. This is good practice: continued boosting outside the audible frequency range could lead to audible distortion. The treble boost does not have much effect on mid-frequencies – again, good practice: harmonics will be boosted, rather than the fundamental frequencies. However, the treble and bass full *cuts* do affect the mid frequencies slightly.

The other curves show similar information for the tone controls in less extreme positions. Really, these curves are the more useful ones: it's rare (or should be) that tone controls are used on full cut or boost. Look for the same pattern: the boosts levelling off at the extreme frequencies; the mid-frequencies being little affected.

Curve 1 shows a treble filter. Like many response specifica-tions, the frequency at which it operates, (the *turnover frequency*) is expressed as the point at which the response has fallen by 3dB from the flat response – 8kHz in this example. The *slope* of the curve describes the rate at which the filter cuts the response. It is given in terms of the number of dBs the response falls for each *doubling* (for a high filter; halving for a low filter) of frequency. In this example, the response between 5kHz and 10kHz has dropped by 6dB, so the slope at this point is *minus 6dB per octave*.

Since the slope varies, the result will vary depending where on the curve it is measured. With perfect filters, the response eventually settles down to a more or less constant value of minus 6dB (per octave), minus 12dB, or minus 18dB depending on the design of the filter. Since everyone would know the shape of the rest of the curve, quoting this *ultimate slope* would be sufficient. But many amplifiers have non-perfect filters, and the ultimate slope is of less help than knowing how the response falls from, say, minus 6dB to minus 20dB. This you can find out only by reading a graph.

Curve 2 shows a high filter with similar slope, but a lower turnover frequency of 5kHz. This will cut high-frequency hiss more, but then it will affect the tonal quality of the music more, too, making it sound duller and lifeless.

The bass filter curves show the effect of different slopes. At around 60Hz, both filters have cut the response by about 8dB. But to do this, curve 3 (which has a less steep slope than curve 4) has to affect frequencies as high as 100Hz – certainly affecting the tonal character of the music more than curve 4 with its faster roll-off and lower turnover frequency.

Complex load

Some loudspeakers are thought to be difficult to drive, and some amplifiers are thought to be better at driving them than others. The problems in matching amplifiers to loudspeakers may be one of the reasons why some listening tests give different results from others, and why traditional technical tests do not predict how amplifiers will sound.

One factor behind all this is that real loudspeakers are a much more complicated load for amplifiers to drive than the simple 8ohm resistors that are usually used in tests. A more realistic test would be to design a typical loudspeaker-type load, and to measure power output and distortion using that. The latest IHF amplifier Standard specifies a suitable complex load, which some reviewers in this country use.

The usual test method is to measure the output across the complex load at the onset of clipping, and to compare this with the output for the same distortion level when the amplifier is driving just a simple resistor. An amplifier unaffected by the type of load it is driving (and which should therefore be capable of driving difficult loudspeakers) will give the same voltage into both loads.

The complex load test can be useful, particularly during the design stages of a new amplifier; but even if an amplifier fails, it may not give audible problems, especially if it is not being driven close to full output power. So don't worry too much about this test.

It is unlikely, anyway, that you would see manufacturers quoting results into a complex load.

Stereo separation

Stereo separation, or crosstalk, is a measure of how well the amplifier keeps apart the left and the right channel signals. If there is any leakage from one channel to the other, the stereo effect will become less clear, and because the leakage may be highly distorted, the overall distortion performance can be poorer, too. Normal distortion measurements, made by sending a signal through one channel only, and measuring only that channel, would fail to pick this up.

Crosstalk measurements are made simply by sending a signal through one channel, and measuring the output on the other; if there is no output, the separation is infinite. In practice, the non-speaking channel on an amplifier may be 50dB down in response (ie, the separation is 50dB). Anything over about 40dB is good for an amplifier, and should not cause any problems; indeed, amplifiers rarely have poor crosstalk measurements. (Tuners and pick-up cartridges are generally the limiting factors in a system.)

STEREO SEPARATION AT DIFFERENT FREQUENCIES

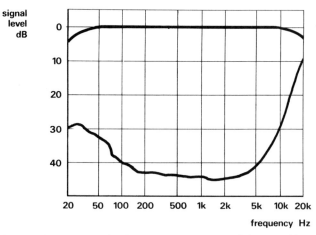

The top line shows the frequency response of the speaking channel – the channel the signal is being sent through. The bottom line is the response from the non-speaking channel. Note that it is lowest – ie the separation is greatest – in the middle frequency range; separation here is about 45dB. The separation is less at low and high frequencies – as little as 20dB at 15kHz. Poor separation at the extremes of the frequency range does not spoil the stereo effect too much. But quoting the smallest separation over the range where good separation *is* important would give a more meaningful result than quoting just the largest separation, or the result at say 1kHz (which is what most manufacturers do). DIN Standards specify a range of 500Hz to 6.3kHz; some experts think that a range of up to 10kHz would be better, particularly for FM tuners where poor separation at high frequencies can give rise to spluttering effects.

Chapter 6   Record decks and cartridges

A record deck's guts is the **turntable** unit – which in turn comprises the *platter*, on which the record sits and which spins the records round, and a *motor* to turn the platter. The term turntable can also include the box – or *plinth* – that houses the platter and motor.

The other essential bit of a record deck is the **pick-up arm**. This carries, at one end, the pick-up *cartridge* – the piece of equipment that traces the groove in the record, and retrieves the information buried in it. Pick-up arms are sometimes called tone arms – a term that doesn't really give you any idea of its function: indeed, the last thing a pick-up arm should do is produce tones of its own.

Most record-playing equipment is in the form of record decks – often, especially at the cheaper end of the market, already fitted with a pick-up cartridge (though this is not necessarily a good thing). A few separate turntable units, complete with plinth, are available – usually at the more expensive end of the range. And there are pick-up arms you can buy separately.

How records and record decks work is shown on the next page. Briefly, the record groove holds a direct physical pattern of the sound wave; the stylus in the pick-up cartridge traces the pattern and vibrates in sympathy; the vibrations are turned by the cartridge into an electrical signal which is passed to the amplifier.

The problems with record decks

A few years ago, many people were saying that record decks could hardly be improved further. The basic tests – such as those for wow and flutter – gave good results, and even cartridges (once notoriously the weak point in the chain) were giving reasonable frequency response and distortion performance. Then, as with amplifiers, some reviewers started to question this. They claimed to be able to hear differences between record decks – and were able to call on relatively simple mechanical theories to back up their claims. What are the problem areas, and how likely is it that they produce audible differences between decks that otherwise measure similarly?

Alignment is the first problem, especially as *Which?* tests have shown that you cannot rely on a deck, even one with the cartridge already fitted, being accurately set up when you buy it. Indeed, you cannot always rely on the manufacturer's instructions for setting up being complete or totally accurate, either.

A good dealer could do a lot of the setting up for you in the shop – but there is some fine tuning that can be done only when the deck is in position. So, unless you are buying an expensive outfit and can persuade the dealer to set it up in your home, you will have to carry out at least some of the alignments yourself – and if you can't find a good dealer, you will probably have to do it all. It doesn't usually involve great skill, but it needs a lot of patience; details about what is involved start on page 71.

Good pick-up cartridge response is the next issue. This is partly a matter of getting a good cartridge, of course, but it is also a matter of matching the characteristics of cartridge and amplifier. The electrical properties of the cartridge and the pick-up input of the amplifier can interact in a way that will alter the frequency response (for better or for worse). The effect can be quite noticeable – and is almost certainly one reason why amplifiers appear to sound different from each other. It is often easy to correct a mis-match between amplifier and cartridge, though unfortunately information on which cartridge and amplifier combinations are good and bad is not easy to find. More details about this problem and how to solve it on page 69.

CHAPTER 6

HOW RECORD GROOVES WORK

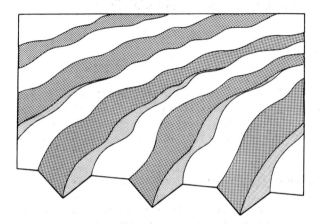

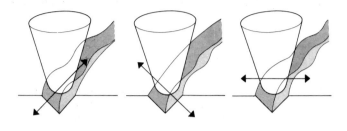

A record has a *single* spiral groove cut on each side. (It is a useful shorthand to describe parts of the groove that lie side-by-side – ie one revolution apart – as adjacent grooves.)

In the recorded area, the groove is shaped into a waveform – in fact, a direct representation of a sound wave (that is, the greater the size of the wiggle in the groove, the louder is the sound being represented; the faster the wiggle changes, the greater the frequency).

The groove could be wiggled (or *modulated*) in any direction – for example, from side to side across the surface of the record, or up and down through the thickness of the record. Very early records were cut with up-and-down modulation (called *vertical*, or hill and dale, cutting), but the record industry soon standardised on side-to-side, or *lateral*, modulation for its **mono** records.

The cutter is wedge-shaped, so the width of the groove in a vertical cut varies, depending on the loudness of the sound being recorded (the louder the sound, the wider the top of the groove; see diagram of magnified grooves, *above*). Whatever the shape of the cutter, there is obviously a limit to the loudness of sound that can be recorded with a vertical cut: the louder the sound, the nearer the centre of the thickness of the record the cut will come.

With a lateral cut, the loudness of the sound does not affect the groove width. Nor is there the same basic restriction on the loudness of sound that can be cut – so long as the distance between adjacent grooves (the *pitch*) is sufficient to accommodate the amount of modulation needed. If the pitch is too fine, of course, one groove will meet the next.

To make the best use of record space, many records are cut with variable pitch: the space between adjacent grooves is automatically widened when a loud sound is to be cut.

Stereo records must have two separate sections of information – left channel and right channel. And they must be playable on mono equipment (in mono). The solution is for each channel to have a combination of vertical and lateral modulation. Each of the groove walls is tilted at an angle of 45°, and thus at an angle of 90° to each other. If the left-hand wall only is modulated, the stylus will move as shown in the *left-hand diagram above* and it will interpret the movement as left channel only; if the right-hand wall only is modulated, the stylus moves as shown in the *middle diagram above* – right channel only.

A mono signal is equivalent to equal modulation on both left and right channels (or groove walls) simultaneously – the *right-hand diagram above* shows that the stylus is forced to move only from side to side. Since it is only lateral movement that a mono cartridge will react to, the system is stereo/mono compatible.

In practice, at any point on the groove there will be a bit of left and a bit of right information, with the proportions continually changing. So the angle the stylus wiggles at will be constantly changing, and the modulation will be different on each groove wall. But there will almost always be some lateral movement, which represents the mono information, and some vertical movement, which represents the stereo information.

Avoiding unwanted vibrations throughout the system is the next problem. Unwanted vibrations cause the stylus or cartridge to move when it shouldn't. Since the system has little to help it distinguish vibrations of the stylus that are due to music in the record groove from other, unwanted, vibrations, the resulting sound can be different from what it should be. There are two main types of unwanted vibration: mechanical ones, such as someone walking across the floor, and acoustic ones – chiefly the sound from the loudspeaker which is reproducing the music the record deck is playing. Any part of the record deck might be guilty of allowing unwanted cartridge movement or of not preventing spurious vibrations from reaching the stylus – the turntable plinth, the pick-up arm, the deck lid, or the table the deck is stood on. Matching the deck to the room to avoid this is explained in detail on page 70.

Avoiding unwanted resonances is the final part of the problem. Resonance happens when vibration of a particular frequency continues and grows uncontrollably. This obviously affects the frequency response, and so can affect sound quality. In a perfect record deck, the only thing that would vibrate at all would be the stylus. But real materials are never perfect, so there will always be some vibration in the plinth, platter, and particularly in the arm. And if there is vibration, there may be resonance. The aim of a record deck designer is to reduce the resonance as much as possible, or to ensure that it happens over a frequency range where it will not affect the sound quality.

One resonance that is quite important to control is that due to the interaction between the springiness of the stylus assembly, and the weight of the pick-up arm. Getting a good match between cartridge and arm to control this is explained on pages 70 and 71.

How much do all these problems matter in practice? Some reviewers will say that they are extremely important (especially the problems of resonance and vibrations), and will claim to hear differences between record decks that are as large, if not larger, than those between any other parts of a hi-fi system. They will also point to technical tests that are able to show up differences in the resonance behaviour of different decks or arms. And they advance the theory that, as the record deck is at the beginning of the hi-fi chain, it is therefore the most important component: the amplifier and loudspeaker can do nothing to improve on the quality of the signal the record deck feeds them.

On the other hand, the record deck is *not* actually the beginning of the chain. The original recording and the disc which is subsequently made from it are really the beginning. And in many cases, records are of lower quality than the systems on which they are played. *Which?* listening tests have shown that it is fairly difficult to hear differences between many good quality record decks (at least, in the price range that *Which?* reports on) and that most of the differences are due to basic reasons such as wow and flutter, or poor frequency response, rather than anything more esoteric, like uncontrolled resonances. The results of these tests also indicate that if resonances and vibrations do upset the sound quality, then the extent is likely to be influenced by the room, and the layout of the equipment in it, at least as much as by the brand of record deck. This also means that the results of resonance tests can be only a pointer to the behaviour of record decks.

So our advice is this. For most people, the important things about a record deck is that it produces inaudible amounts of wow and flutter and rumble (or at least, amounts that are lower than those found on real records), and that the cartridge has a good frequency response and distortion performance (when connected to the amplifier you intend to use). It is also fairly important that the cartridge matches the arm tolerably well. *Which?* **tests have shown that this combination is not difficult to find, even in relatively inexpensive decks and cartridges. But to get the best results (the sort of results that** *Which?* **gets in its tests) it is important to pay great attention to setting up the deck properly and to matching the cartridge and amplifier, at least some attention to matching cartridge and arm, and some to the layout of equipment in the listening room. All that is dealt with between pages 69 and 74.**

CHAPTER 6

RIAA CORRECTION

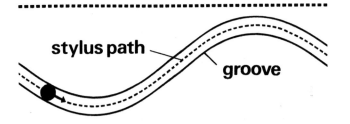

low frequency

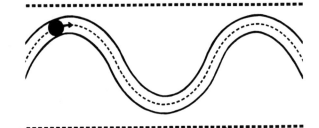

high frequency

For a given loudness of sound, the size of the wiggle recorded is more or less the same, whatever the frequency of the sound. This *constant amplitude recording* may seem an obvious thing to do, but the reasons for it are not as simple as they may appear, and it does bring problems.

As the size of the groove wiggle is the same whatever the frequency, the stylus velocity has to vary with frequency. The diagram above shows two recorded waves which have the same amplitude (the distance from peak to valley is the same), but different frequencies (in the same length of groove, the higher frequency sound has more peaks). The time taken for the stylus to trace each length of groove is the same: it depends on the speed at which the record rotates, and the distance each groove is from the centre of the record – both the same in this example. So for the higher-frequency sound, the stylus has to move along a longer path than for the lower-frequency sound, but in the same time. So it has to travel faster – that is, its velocity has to be higher.

This gives two problems. First, at very high frequencies, the velocity may be too high for the stylus to cope with. Second, with magnetic pick-up cartridges – virtually the only type capable of giving hi-fi performance – the output signal depends not on the amplitude of the stylus wiggle, but on its velocity. So, although the frequency of the output signal is the same as the frequency of the recorded groove, its amplitude does not follow the amplitude of the recorded signal.

However, this method of recording has some advantages. It maintains a good signal to noise ratio at high frequencies, and it ensures that the groove pitch does not become excessively wide at low frequencies. And the problems it causes can be fairly easily solved.

It is fairly easy to design an electrical circuit to correct for the second problem, and since all record manufacturers use a standard method of recording, a standard correction circuit will do for all records, and can be built into each amplifier. This is known as *RIAA correction* (after the Record Industries Association of America), and the magnetic pick-up input on an amplifier is often called the RIAA input.

Getting over the problem of excessively high velocities at high frequencies is more a matter of luck than judgement: high-frequency sounds in music are usually at a low level (because they tend to be harmonics rather than fundamentals), so the velocities are not as large as they might be.

Since velocity is just as important as amplitude, standard recording levels on records are often given in terms of the velocity that the stylus will have to move at to trace the groove – for example, 3.54 or 5cm per sec are two widely-used velocities on test records.

The maximum velocity depends on the record, and on the type of music being recorded: so-called *supercut* discs can give velocities as high as 50cm/sec at very high frequencies.

Pick-up cartridge outputs are also given in terms of the output voltage for a particular velocity. Most moving-coil magnetic cartridges give outputs around 1mV rms for a velocity of 1cm/sec (written as 1mV/cm/sec). Some might go as high as 2mV/cm/sec. So on a 50cm/sec supercut disc, the amplifier's pick-up input must be able to handle at least 100mV at high frequencies without overloading and causing distortion. Amplifiers are not seriously poorer than this, so it's rarely a problem.

FEATURES ○|○|○|○|○|

There is little in the way of knobs and switches on record decks – and most of what is available is irrelevant to sound quality (and often of little extra convenience).

Pick-up cartridge and stylus

The stylus is the really fundamental bit of a record deck: the rest of the deck is there only to allow the stylus to trace out the record groove in a controlled way.

The stylus itself is usually connected to a *cantilever arm* (usually bought together). The cantilever transmits the wiggles from the stylus to the cartridge body, and in the body there is some means (a *transducer*) of turning the physical motion into an electrical signal. Note that an electrical signal is produced only when there is relative motion between the cartridge body and the stylus; anything that affects the amount of this relative motion can affect sound quality (see *Matching*, page 69).

Overwhelmingly, hi-fi cartridges make use of a magnetic transducer: moving a magnet near a coil of wire (or vice versa) will cause a current of electricity to flow in the coil.

Most hi-fi cartridges are **moving-magnet** ones – the cantilever has tiny permanent magnets attached to its end, and these move between coils fixed in the cartridge body. Even though the magnets and coils have to be extremely small, this type of cartridge is reasonably easy to make, gives quite a large output signal, and can be made to trace the groove well.

The main alternative to the moving-magnet cartridge is the **moving-coil**. The cantilever has tiny electrical coils connected to it, and these move between fixed permanent magnets. Moving-coil cartridges are becoming more and more popular among enthusiasts. They do have, however, many problems. They tend to be much more expensive; the output is usually very low – so, unless your amplifier has a moving-coil input, you may need an additional pre-preamplifier or head amplifier; they usually also have poorer tracking ability, which might lead to greater distortion on peaks.

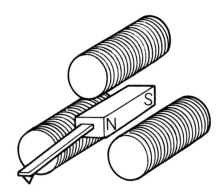

Simplified view of a moving-magnet cartridge. The electrical output signal is generated according to the relative movement between the coils and the magnet. The record groove carries a direct physical pattern of the sound wave – so for the cartridge output to be a direct electrical pattern, the whole of the stylus wiggle is wanted, and the cartridge must appear completely stationary to the stylus. For example, if the cartridge moved as much as the stylus, and in the same direction, the wiggle of the stylus relative to the cartridge body would be zero, and there would be no signal output from the cartridge. Any other movement of the cartridge would give an output different from the one given when the cartridge is stationary – which is another way of saying that the output would be distorted.

On the other hand, there are movements of the stylus that we do not want to pick up – for example, the inevitable warps in the record surface will make the stylus move up and down, and if this gives an output signal, then reproduction will be marred (the warp will give a very low frequency output which would not be heard directly, but which could cause intermodulation distortion). In this case, we would want the cartridge to move in step with, and by the same amount as, the stylus movement. The continual spiralling of the groove towards the centre of the record will also cause the stylus to move; this is another movement which must not result in an output from the cartridge, so again the cartridge body must move exactly in step with the stylus.

To sum up: *the cartridge body must appear perfectly stationary to the stylus when the stylus movement is due to modulation in the record groove; otherwise it must move perfectly in step with the stylus movement.*

Similarly, the stylus itself should move only when a sound pattern in the groove makes it – any other vibrations which might move the stylus must be ignored.

CHAPTER 6

GROOVES & STYLUSES

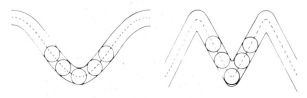

The record cutter is a sharp-edged tool and when fed with a sine-wave signal, will cut a perfect sine wave in the record. So long as the frequency is relatively low, the rounded stylus can trace the grooves accurately (*above left*). But at very high frequencies (*above right*), the stylus cannot fit into the sharp peaks of the groove, and the path it follows is a distorted sine wave – leading to a type of distorted sound called **tracing distortion**.

The simplest way of getting the stylus to trace high frequency grooves properly is to reduce its diameter (*left*). Then it will fit into the peaks better. But too small a stylus has many disadvantages. First, it has to be at least big enough to make contact with both groove walls – or else it will flop about in the groove, causing unwanted signals; second, even this big is not big enough – there is always dirt in the bottom of the groove which the stylus must ride clear of, if the signal is not to be noisy. Third, the smaller the area of stylus in contact with the groove wall, the greater the tendency for the stylus to dig into the walls, causing record wear. The only solution to this is to reduce the playing weight; but the lower the playing weight the greater the tendency for the stylus to loose secure contact with the groove, leading to distortion.

The main solution to the problems is to use an **elliptical stylus** (*centre* and, viewed from above, *right*). These are wide from side to side, narrow from front to back. The narrow dimension fits into peaks, and the wide dimension holds the stylus firmly against the walls, and clear of the bottom of the groove.

Stylus types. The stylus which traces the record groove is not the same shape as the cutter which formed the groove in the first place. The consequence of using a different shape is that the stylus does not follow exactly the same path as the original cutter. This error leads to a form of distortion called *tracing distortion*; it is worst at the centre of the record and for high frequencies, where the wiggles in the groove are very tight. The diagrams alongside illustrate this problem.

The basic stylus shape is **spherical**, and the smaller its diameter, the more readily it will trace the groove. But there are limits to how small it can be made: too small, and it will not only start to chisel out its own groove, but it will rub along the bottom of the groove (which is usually full of dirt) giving a noisy signal and loose contact with the walls. The answer is to use a **bi-radial** or **elliptical** stylus: its large width keeps it clear of the groove's bottom, and in contact with the walls, and its small length enables it to trace the grooves more accurately. The only problem is that it has to be accurately mounted on to the stylus – if it is twisted, the benefits are lost. And quality control on many cartridges is not all that good.

There are other, more complicated, shapes which are claimed to fit the groove better, and give better performance – the so-called **extended** or **line contact** shapes. Manufacturer's names for these vary – *Aliptic, Hyper elliptical, Stereohedron* and *Super elliptical*, for example – but they all have one thing in common: their alignment is even more critical than with a bi-radial stylus – more vulnerable to poor quality control.

Another term you may come across is a *nude* stylus. This simply means that the stone forming the stylus tip (usually a diamond) is mounted directly on to the cantilever. A stylus which is fully dressed, so to speak, has only its tip formed of diamond and this is bonded to a (usually) metal tube attached to the cantilever. Neither method of mounting is inherently superior.

Although the stylus is a small part of the cartridge, its **replacement cost** is not insignificant. On cartridges costing up to about £15, a new stylus is often only a couple of pounds cheaper than the cartridge itself. With more expensive cartridges, expect to pay at least *half* as much for a replacement stylus: this rule of thumb holds even with cartridges costing £100 or more. Many moving-coil cartridges have to be sent back to the manufacturer or importer for stylus replacement.

How long should a stylus last? It used to be thought that 1,000 hours (say 1,500 LP sides) was a sensible limit – but if you are a critical listener, you might notice wear in as little as 400 hours' use. So if you listen to records a lot and you use an expensive cartridge, then this running cost can be quite high.

Surprisingly good results can be had from fairly inexpensive cartridges – so before you spend a lot of money on an expensive one, be sure that you really are getting value for money (both in initial and running costs). Unless you can tell differences between moving-magnet and moving-coil sounds, you would be best advised to stick to the more usual moving-magnet types. Do not get beguiled by stylus shapes more esoteric than, say, elliptical: though they can improve some areas of performance, the cartridge has to be inherently very good – and the quality control outstanding – before their virtues are noticeable.

Cartridge sound (particularly its frequency response) can be affected markedly by various matching problems: so before you blame a cartridge for a sound you don't like, check that the fault does not lie in incorrect matching.

Pick-up arms

There are relatively few separate arms on sale; many of them are sophisticated designs costing a couple of hundred pounds or more. As with decks, it is unlikely that for many people the choice of arm itself will make a lot of difference to sound quality (though choosing a matched arm and cartridge might).

Most arms have a *pivot* near one end, so that the stylus moves in an arc over the centre of the record. The cutter which cuts the record, however, moves in a straight line over the record. Because the two paths are different, some distortion will be generated on replay. The usual way of minimising this is to use geometrical tricks – putting the stylus at an angle to the arm, and arranging for the stylus to describe an arc beyond the centre of the record. See *Tracking error* (page 72) for more details.

Manufacturers use different methods of arranging this geometry. Some use a *straight pick-up arm* tube, with the cartridge headshell (the bit of the arm the cartridge is fastened to) set at an angle to it. Others use shaped arm tubes – *S shapes* are particularly popular. On the one hand, the straight-tube approach has some advantages because it minimises the amount of material in the arm, and hence its weight; on the other hand, an S shape will provide a balancing force to counteract the weight of the offset headshell, which might otherwise tilt the arm and stylus, and cause distortion. A *longer-than-usual arm* will also reduce the amount of arc – but long arms are heavier and not now fashionable.

With pivot arms there are various forms of *bearing* at the pivot end. The main thing about the bearing is that it should produce low levels of friction, so that it can move over the record surface and up and down over record warps without affecting the movement of the stylus at all. How free the bearings have to be for this depends partly on the stylus: the less force required to move the stylus (the higher its *compliance*), the lower the pivot friction must be.

However, and as usual, no one type of bearing is necessarily better for friction than any other.

The occasional truly **radial-tracking arm** appears. One approach is to use a system of rods and levers on an ordinary pivoting arm to constantly adjust the angle of the headshell so that the stylus is kept parallel to the record grooves. The snag is that all the extra mechanical bits may affect the arm's performance more than it helps tracking distortion. Another approach is to use a **parallel tracking-arm**, and slide the whole arm along a track, so that the stylus moves along a straight line. The problem here is that the arm cannot be guided over the record surface by the record groove, as it can with a pivoted arm. Instead, a complex motor system has to be used with sensors to ensure that the cartridge remains in the right groove: decks with parallel tracking arms tend to be expensive.

The **effective arm mass** is a measure of how much

resistance the arm puts up to the stylus moving up and down. It is not quite the same as the weight of the arm, but takes into account things like how the weight is distributed along the length of the arm. For the very best sound quality, it is important to match effective arm mass and cartridge compliance (see page 71 for more details). This means that a low effective mass is not always the best thing – but it happens that many cartridges do have quite high compliance, and in rough terms, high compliance cartridges, especially moving-magnet ones, can give better performance than low compliance ones, so a low mass arm is often the one to go for.

The **headshell** can contribute quite a lot to the effective mass, because it is at the end of the arm. So, much effort has been put into making its mass as small as possible – drilling holes in it; making it of esoteric lightweight materials; making it as small as possible; even dispensing with it altogether and mounting the cartridge directly on to the arm. Some of these attempts can lead to lack of rigidity and unwanted resonances – which in turn are supposed to lead to poorer sound quality. This is one reason why purists prefer non-detachable headshells: they increase rigidity, and reduce arm mass.

An enormous fuss is made about tracking error in pick-up arms – possibly because the geometrical problems are easy to understand. But *Which?* tests have shown that it is possible, with careful setting up, to adjust tracking error quite satisfactorily on most decks. Parallel tracking arms are unlikely to sound any better than a simple pivot arm. Most people, too, are unlikely to notice the difference in sound quality between one arm and another – as long as the cartridge used in it is reasonably well matched to the arm.

Types of drive

One way of rotating the turntable is to have it driven by a high-speed electric motor via a rubber belt – **belt drive**. The belt passes round a small pulley on the shaft of the motor, and a much larger pulley on the turntable platter – perhaps the rim of the platter itself – to gear down the high speed motor to the correct speed for playing re-

cords. To cope with speeds other than 33⅓rpm, the motor shaft might have extra pulleys of different diameters; simple levers connected to a speed-change control haul the belt from one pulley to another.

Simple though this system is, it does have advantages. With decent engineering, the turntable can be made to rotate at the correct speed, and any slight unevenness in the motor can be ironed out by the flexible rubber belt before it gets to the turntable, thus reducing wow and flutter.

The other main system is **direct drive**. The platter forms part of the motor itself, and the motor rotates at the same speed as the platter. So there's no flexible belt to iron out small speed instabilities; an electronic *servo system* has to be used instead. The servo senses when the speed of the motor changes from what it should be, and corrects it instantly (more or less) – speeding up the motor if it seems to be running slowly; or slowing it down if it seems to be running fast. The problem is the more or less: a servo is good at ironing out relatively long-term speed drift, but not so good where the speed change happens very quickly. Often the best method of control-

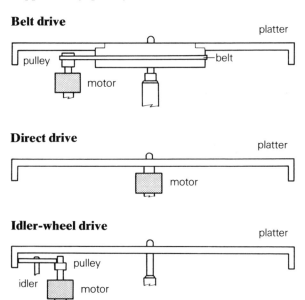

Belt drive

Direct drive

Idler-wheel drive

ling these transient speed changes is to design them out in the first place – by using a very heavy platter, whose speed is difficult to change quickly unless a very large force is applied.

Because adding anything 'electronic' to a piece of equipment (whether it's a car, washing machine, or hi-fi) is usually reckoned a good way of increasing sales, some manufacturers have taken to adding electronic servo control to belt-drive decks. But a well-engineered belt drive deck should not need servo control at all.

The third main type of drive – **idler-wheel** – was popular in hi-fi decks some years ago, and is still to be found in some cheaper equipment. Here, a fast-running motor with a capstan spindle instead of a pulley is connected to the platter by a rubber wheel which presses against the capstan of the motor and the rim of the platter. When a very heavy platter is used, most areas of performance are good – but rumble (see page 75) is often a problem. In cheap decks, which are likely to be used with loudspeakers having a restricted bass frequency response, rumble does not matter so much.

On the whole, therefore, idler-wheel drive is not the best choice for a high-quality deck. Purists still swear by belt-drive, but nowadays, there's no reason to think that a direct-drive deck is necessarily inferior.

Automation

There is an amount of mechanical to-ing and fro-ing on a record deck: the turntable has to start rotating; the pick-up arm has to leave its rest, and land gently on the record at the correct place; when the record has finished playing the pick-up arm has to lift, return to its rest, and the turntable switch off.

With many decks, particularly the more expensive ones, the only help you are likely to have with all this is a *cueing lever*. Operating the lever lifts a small support underneath the pick-up arm, pushing the free end into the air off its rest, or off the record. The mechanism is usually a simple set of mechanical levers, though manufacturers have been known to go so far as fit a separate little motor to do it. You then move the arm across the record, so that the stylus is above the point you want to start at (or the arm is back over its rest), and operate the cueing lever again, to lower the arm. Decks whose only automation is a cueing lever are called **manual** decks.

A good cueing lever will lower the arm gently – but not maddeningly slowly – and with no sideways drift, so that the stylus drops on to the groove you positioned it above. If you can't reliably hit the space between tracks, for example, it's not a good cueing device.

At the other end of the scale, a few decks will do all the hard work for you. The turntable will start from rest; the pick up arm will put itself on the record (at the beginning of a 12-inch LP); at the end of the record, the pick-up arm will return to its rest; the turntable will switch itself off. These are often called **fully-automatic** decks. Some purists frown on automatic decks as many of the mechanisms needed to move the pick-up arm on and off the record, and to sense when the arm has come to the end of the record, hinder the free movement of the arm; in turn, this prevents the use of delicate, and possibly higher-quality, cartridges. But modern mechanisms seem to be better, and *Which?* tests have demonstrated that automatic decks can take quite delicate – and certainly good-quality – cartridges.

Between these two extremes, there is a range of **semi-automatic** decks. Many don't place the stylus at the beginning of the record, or return the pick-up arm to rest.

If you want what some people think of as automation – the ability to play a stack of records one after the other – you need an **auto-changer**. Auto-changers are clearly clever enough – some can even decide whether the next record to be played is an LP or a 45rpm single, and change speed and position the pick-up arm accordingly. But they have no place in a hi-fi system: leaving precious records out of their sleeves, placed horizontally, and supported only by a thin centre spindle is bad enough (see *Record care*, page 78) – but then dropping them on to another already-revolving record seems almost criminal. There are no *hi-fi* auto-changers.

The occasional **micro-processor** controlled deck is available. They can be programmed to play particular tracks in a particular order, and as many times as you like. Even if you wanted to listen to odd tracks or excerpts this way, you'd have to want it very badly, as these decks are often very expensive.

One or two **add-on** devices are available for manual decks, which lift the arm clear of the record surface at the end. This at least stops the annoying noise of the stylus tracing the locked run-out groove, and may cut down on stylus wear too. These arm lifts are not connected to the arm at all, so they cannot upset its performance. They are worth thinking about if you have a manual deck.

Automation is helpful and convenient, and a degree of it need not compromise a deck's performance. If you want automation, check carefully exactly what the deck does before you buy – functions vary. Relatively few decks, however, have much automatic control – you will have a wider choice, particularly among the more expensive decks, if you stick to manual operation (and invest, perhaps, in an add-on arm-lifter).

Speed controls

Most decks offer two speeds – *33⅓rpm* and *45rpm*. A couple of decks offer only 33⅓rpm on the grounds that speed-change mechanisms must compromise performance, and any cheap old record player will do to play pop singles. But record manufacturers have confounded them by turning out the occasional 45rpm 12-inch disc, using the higher speed specifically to get better performance.

One or two decks offer *78rpm* as well; you will need this only if you have a collection of old 78s. Since 78s need a different stylus, and since they do not demand the highest of hi-fi, you may do best to buy a cheap deck or record player with 78rpm just for your 78s, leaving your choice of hi-fi deck much less restricted.

On some decks you can **vary the speed** a little either side of the nominal speed, and some manufacturers make great play in their advertising about the speed accuracy of their products – particularly if they have electronic speed control. But absolute speed accuracy is no real virtue: unless you are blessed (or cursed) with perfect pitch, you would probably find it difficult to tell if your music was being played to within a semitone of the correct pitch (unless you were comparing it with another deck at the same time). A semitone is a difference in pitch (and therefore in turntable speed) of *six per cent*; any deck – even fairly cheap low-fi ones – are unlikely to have a speed error of more than *one per cent* or so.

A variable speed control is useful if you *have* got perfect pitch, or if you want to play a fixed-pitch musical instrument (like a piano) along with a record. Fixed-speed decks, even if highly accurate, are no use here: the recording itself may not be perfectly in tune.

Decks with variable controls usually have a **stroboscope** to help you set the speed right. This is a disc with marks – see diagram below – which appear stationary under electric light when the speed is right. They are strictly unnecessary – since what is important is whether *you* think the speed is right or not. But if a deck has a strobe, it is better if it is fixed to the *rim* of the platter rather than the top. That way, it is visible when a record is actually being played – some decks show slight differences in speed when loaded with a record or a cleaning device. Unfortunately, this means you will be able to see the strobe when you should be concentrating on the music – and unless you exert some will-power, you might find yourself leaping up and down to correct small changes in speed that you can only see, not hear. If the mains frequency changes (it can do, up to a maximum of two per cent) a strobe can give a false reading anyway.

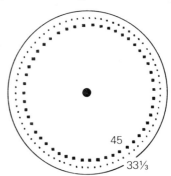

A stroboscope consists of markings around the circumference of a circle. Under an ordinary mains electric light bulb, these will appear stationary if the turntable is moving at the correct speed. If it is moving slightly too fast or too slow, then the marks will appear to be drifting slightly one way or the other; if the speed is way out, the marks will appear as a blur (as they will do if you view them in natural light).

Absolute speed accuracy is not very important, despite what the copywriters might say, and any fixed-speed deck should be accurate enough for most people. If you have perfect pitch, or you want to play say, a piano along with records, you need a variable-speed deck, rather than a highly-accurate fixed-speed one. With a variable speed deck, leave the controls strictly alone unless you can *hear* problems of pitch.

Types of chassis

The construction of the plinth, and the method of mounting the motor and turntable unit to it, are important in reducing the level of unwanted vibrations. But the particular technique used is generally less important than the way it is applied.

A **suspended sub-chassis** – where the turntable, motor and arm are all mounted rigidly together, and hung from the plinth on springs – is thought by many experts to be a good method of reducing vibration. But if it is poorly designed, it can still suffer from resonances.

Some manufacturers make great play of using *inert materials* in their plinths, or very heavy materials – but these won't necessarily damp out the crucial vibrations.

Do not place too much trust in the technique used in a particular chassis – it may still not give good results if the engineering is poor. In any case, *Which?* tests indicate that vibration performance might have more to do with the room layout than with the suspension system used in the deck. See *Setting up* (page 71) for hints on how to get the best from your deck.

Matching

A record deck system consists of various components. For best results, each component must perform well – but also, each must be compatible with the others. A particular cartridge and arm, for example, may each perform well in isolation, but the combination may not. There are three areas where, to get the best results, careful matching of the different components may be required:
- matching cartridge and amplifier
- matching the complete record deck to the room
- matching cartridge and arm.

Cartridge/amplifier matching. Pick-up cartridges are designed to give their flattest frequency response when they are connected to a given electrical load. The load a cartridge sees is the input impedance of the pick-up input of the amplifier it is connected to, and the combined capacitance of the wiring in the pick-up arm and of the amplifier pick-up input. The resistance of most amplifier inputs is 47kohms (or 50kohms), and their capacitance is often around 50pf (picofarads); arm wiring typically adds another 100pf. So most moving-magnet cartridges are designed to give their best results with a loading of 47kohms and 150pf.

But arm wiring and input capacitance can vary, so cartridges should really be tolerant to capacitance variations between say 150pf and 250pf. Some cartridges cannot do this – especially those which have an *inductance* of greater than about 300mH (millihenries) – and if faced with too high an input capacitance, their treble response will be poor, giving a dull sound. On the other hand, some makes of cartridge are notorious for needing excess *capacitance* – perhaps as high as 450pf. Load resistance will also affect the frequency response.

Manufacturers sometimes publish recommended loadings for their cartridges – but reviewers do not always agree with them. Amplifier manufacturers rarely quote input capacitance. (They do quote input impedance, but as it is almost always 47kohm, this is not much help.)

As usual, the better reviews (including *Which?*) will indicate if a cartridge's requirements or an amplifier's offering, are at all unusual. But unless you have been warned that a bright or dull sound is due to cartridge/amplifier mismatch, it is often difficult to pin this down as the source of a problem in a complete system; the fault may be with the cartridge itself, or with the loudspeakers.

There is nothing wrong in trying to correct a poor frequency response which may be, say, a loudspeaker problem by altering the cartridge/amplifier matching. But it is sometimes difficult to get two wrongs to make a right, and the resulting frequency response may sound as bad – albeit in different ways – as it did originally. It is better, if possible, to ensure that each component or interconnection provides a flat frequency response, than to rely on being able to correct a fault in one part of the chain by introducing a complementary fault in another part.

How important is this matching? It seems as though it could have a large effect. Reviewers who have carried out serious listening tests suggest that the greater part of any difference in sound quality among amplifiers can be attributed to differences in their pick-up input capacitance. *Which?* tests tend to confirm this. So, if you are concerned by differences that you hear between amplifiers, it would be very sensible to check that they are not due just to differences in matching.

Altering input resistance and capacitance is fairly easy. Some amplifiers are equipped with switchable loadings (though often, just resistance), and there are a number of equalisers on sale, that simply plug in to the back of the amplifier. Another, perhaps less elegant, way of altering capacitance is to change the length of the lead connecting the record deck to the amplifier. Shortening it will reduce the capacitance; lengthening will increase it.

Deciding by how much to alter the capacitance is rather more difficult. If you have access to specifications, review data and so on, then you can use that. Otherwise you will have to carry out listening tests to find out what the effects of changing capacitance are.

Moving-coil cartridges, because they generally have low inductance, are much less susceptible to changes in input loading; and so are many moving-magnet cartridges with low inductance. On the whole, this is a good feature: it gives you one less matching problem to worry about. And it means that, when you are looking for the source of frequency response errors, one area has been eliminated.

Record deck/room matching. A record-playing system works because the pick-up stylus vibrates. The only vibrations we want the stylus to make are those caused by modulations in the record groove. But all sorts of other vibrations can move the stylus: people walking across the room, for example; even acoustic vibrations from the loudspeakers that the stylus is eventually feeding. Gross vibrations – people jumping up and down, for example – can bounce the stylus out of its groove, but gentler vibrations can have a more subtle effect on sound quality, perhaps making it sound muddy.

Some research has been done into how important acoustic and vibrational effects are on sound quality. One of the problems is that the effect is likely to vary from room to room, so it is difficult to correlate the results of technical and listening tests. *Which?* listening tests, too, seem to suggest that attention to room effects and record deck mounting are more important than looking for a deck that's resistant to vibration.

Make sure that your deck is mounted as well as possible before you consider changing it for one with allegedly better performance. The main rule is to mount the deck on a heavy slab, preferably fixed to the wall, especially if the floor is a suspended wooden one.

Apart from that, the only sensible, if limited, advice is to experiment for yourself. You may get better results with the deck mounted out of the line of fire of the speakers, which generally means as far away as possible. This is hardly the most convenient position; mounted close to where you sit while listening obviously makes record changing much easier. So, even if you do hear a difference between those two positions, you might be prepared to sacrifice a small amount of sound quality for greater convenience. Similarly, a particular deck might sound better with the lid raised (on the other hand, it might sound better with the lid closed). But a closed lid protects the record better – which again may be worth more than a small improvement in sound quality.

Cartridge/arm matching. Cartridges and arms form a mechanical system with some interesting properties. When the stylus moves at very low frequencies, the cartridge will tend to move in step with it – the *stiffness region*. When the stylus moves at high frequencies, the cartridge will tend to remain stationary – the *compliance region*. The frequency range in between these two areas is called the *region of resonance*: here, the cartridge can make wild, uncontrollable movements.

This mechanical effect can be very useful. We want the cartridge to appear to be perfectly stationary to the stylus when the movement is due to modulation in the groove, which it will do if the compliance region covers these frequencies. And it should be perfectly movable in step with the stylus movement otherwise, which it will do if the unwanted movements fall into the stiffness region. The only problem is to arrange for the stiffness and compliance regions to cover the right range of frequencies; and the only fly in the ointment is the region of resonance – for violent, uncontrollable motions are certainly not wanted at any frequency.

Since audio signals start at about 20Hz, the compliance region must also start here. Most warps and other record faults generate frequencies of 10Hz and lower – so the stiffness region must cover these frequencies. This leaves a range of 10Hz to 20Hz for the resonance region.

Two factors govern where the resonance region will fall – the compliance of the cartridge, and the weight that the cartridge sees. That weight is not only the static playing weight but also the inertia of the whole arm to movement. This involves the way the weight is distributed over the arm, the freedom of the arm bearings and so on. The weight that the cartridge sees is called the arm's *effective mass* – and together, the cartridge compliance and the arm's effective mass plus the weight of the cartridge determine the resonance frequency of the cartridge/arm system.

So, matching cartridge and arm primarily means selecting a cartridge with a compliance suitable for the arm's effective mass and the weight of the cartridge, to bring the resonance frequency to around 10Hz to 15Hz. As usual, manufacturers' specifications are of little help: cartridge specifications usually give cartridge weight and compliance (expressed in *compliance units*, or cu) – partly because there is little other information they can give, and partly because some people think – wrongly – that a higher compliance automatically means better sound quality. However, you will rarely find arm effective masses being quoted.

How important is this matching? A gross mis-match could bring your record deck to its knees – but this is unlikely to be because of any esoteric resonance effects. It is clear that quite a degree of mis-match can be tolerated. On the basis of one review of arms and cartridges, it seems that most popular cartridges have compliances that are far too high for use in most popular record decks. But the majority of hi-fi users seem happy with the mis-match combinations that they are using.

On the other hand, purists will argue that it is an important factor in getting good sound from a record-playing system; and if you can make a good match, using a cartridge you like the sound of anyway, it makes sense to do so.

Where can you get the information on matching? Some magazine reviews on decks and cartridges will at least guide you in the direction of some acceptable matches – but the advice is generally not very sophisticated. *Which?* tests on record decks give an idea of arm effective masses, and what range of compliance values they are suitable for. And together, the *Hi-fi Choice* books on cartridges and turntables should help you pick an exactly-matching combination.

Setting up

The first job when setting up a turntable is often simply **putting it together**. This usually involves just mounting the platter, and perhaps fixing the cartridge in the arm – a fiddly job, involving fine wires and tiny screws, which you may be able to persuade your dealer to do for you.

Where the arm is separate from the deck, this will have to be mounted on to the plinth too. This can involve anything from simply bolting the pivot assembly to a prepared stand on the plinth, to cutting accurately-positioned large holes in the plinth to take the pivot assembly. Some arms then require careful putting together.

With some esoteric designs of arm and deck, purists will say that it is necessary to have an expert set it up: even things like the position of the wires from the arm are supposed to be able to affect the sound quality. But if you are buying this sort of equipment, you will probably want to go to a dealer that operates this sort of service anyway.

Setting up the arm and cartridge really means being able to make **adjustments**, for tracking error, tilt, vertical tracking angle, playing weight, bias and speed.

Tracking error is a measure of how accurately the stylus is parallel to the groove. The usual way of adjusting this is to move the cartridge slightly forwards or backwards in the headshell – effectively slightly changing the length of the arm. One way of describing the length of the arm is in terms of how much longer it is than the distance from the pivot to the centre of the record. This distance is called *overhang*. Accurately setting the overhang should minimise tracking error. *Which?* tests have found, though, that simply following the deck manufacturers' instructions often gives results that are slightly poorer than they need be. The solution is to buy an alignment gauge – less than £1 from a hi-fi shop. The problem here is that not all alignment gauges are accurate, partly because not everyone agrees on what constitutes minimal tracking error. A good gauge is drawn in *Hi-fi Choice: Turntables and tonearms*, which sets tracking error to zero near the beginning and end of the disc.

As well as setting the amount of overhang correctly, the cartridge must be set at the correct angle to the arm – the *offset angle*. This is done by twisting the cartridge in the slots in the headshell. With an alignment gauge, you

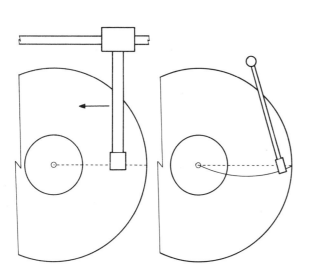

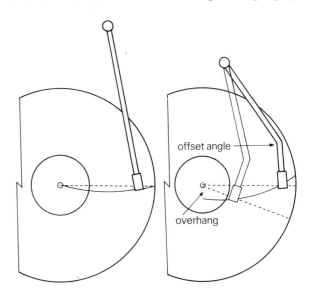

A record cutter (*first picture, going left to right*) is mounted on a radial arm, so it cuts a groove which is always at right angles to a radius.

A pick-up arm (*second picture*) is usually mounted on a pivot, so as it tracks along the groove; the stylus is at right angles to the groove at only one point in its path. Elsewhere there is an error, which leads to distortion.

One solution is a very long arm (*third*): the stylus is still parallel only at one point, but elsewhere the error is not large enough to worry about. However, very long arms bring other problems.

A better solution (*fourth*) is to mount the cartridge at an angle to the arm (the offset angle), and to arrange the arm's pivot point to be near enough to the edge of the turntable so that the stylus traces an arc which ends at the opposite side of the centre of the record from the pivot. The distance by which the arc misses the centre is called the overhang. By correctly adjusting the amount of overhang and offset angle, the tracking error can be made to be zero at two points; at other points, the tracking error can be made negligibly small.

The overhang and offset angle can both be altered by moving the cartridge-fixing bolts in slotted guides in the headshell, or sometimes the whole arm at the pivot end.

A parallel-tracking arm has no tracking error at all.

twist the cartridge until it is parallel with lines drawn on the gauge; with decks having a simple overhang marker, the best you can do is to set the cartridge parallel with the sides of the headshell. This may not be too accurate.

Setting overhang and offset angle are a little fiddly (the overhang really needs to have an accuracy of better than half a millimetre) – but they are worthwhile adjustments to carry out.

Tilt. The stylus point itself should be at exactly 90° to the surface of the record's radius – otherwise it is going to bear more heavily on one side of the groove than the other. Few arms have any real adjustment for tilt, but you may be able to correct gross errors by twisting the headshell in the mounting where it attaches to the arm, or by threading washers on one of the cartridge mounting bolts, between the cartridge and the headshell. A good way of checking the tilt angle is by carefully lowering the stylus on to a record-thickness mirror on the turntable: looking up the pick-up arm from the cartridge end, the stylus and its reflection should be in a straight line.

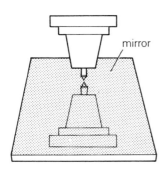

Checking the tilt with a mirror

Vertical tracking angle is the angle that the stylus makes with the record surface. It should point back in the direction the record is rotating from by about 20° (for older records, the correct amount of rake was about 15°). The best way of adjusting this angle is to raise or lower the arm at the pivot.

Many arms do not have a height adjustment, but even on those that do, setting the vertical tracking angle is not easy, because the angle of the stylus cannot be seen readily. The most accurate compromise is to assume that the stylus is fixed at right angles to the cantilever, and to check that the cantilever makes a 20° angle with the record surface. Less satisfactory compromises are to assume that the angle will be right either when the cartridge is horizontal, or when the arm is horizontal – neither very likely.

If there is no arm height adjustment, the vertical tracking angle will have to be altered by putting small spacers between the cartridge body and the headshell – at the front of the cartridge to increase the angle, at the rear to decrease it.

A highly-accurate vertical tracking angle is not necessary, so if your arm has no adjustment, or you find it too difficult to adjust, it is probably best to leave it alone.

If you do adjust the vertical tracking angle, though, it must be done *after* setting the playing weight (see below), since this will affect the angle.

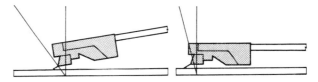

Reducing the vertical tracking angle by lowering the arm at its pivot

Playing weight, sometimes called tracking force, stylus pressure or downforce, is the most important adjustment. It is the force with which the stylus presses down on the record groove. Too light a weight, and the stylus will not be in secure contact with the walls of the groove. At best, this may cause distortion; at worst, it may cause the stylus to jump in the groove, perhaps damaging the record. Too heavy a playing weight can cause records to wear more quickly.(In fact, it is probably *pressure* that determines wear. So the larger contact area of a spherical stylus can track at a higher playing weight than an elliptical one, without causing wear problems.)

CHAPTER 6

The optimum weight varies from cartridge to cartridge. For best results it should be set using a *test disc* (a record with different tones and signals on it, for checking and adjusting cartridge alignment and so on). An easier alternative is to use the cartridge manufacturer's recommendations. Usually a minimum and a maximum playing weight are specified; the best weight is almost always at or near the maximum, and there is nothing to be gained (except distortion and damage) from trying to track at much less than that.

The main problem with setting playing weight is that you have to trust the accuracy of the calibrations on the pick-up arm's playing weight adjuster. Luckily such trust is usually reasonably well-founded, but if you are worried, you could buy a set of tiny weighing scales – a *stylus balance* – to help you (though your money would probably be better spent on a test record).

Before setting playing weight, the arm has to be set into perfect balance so that the ends are tilting neither up nor down. A counterweight, which often forms part of the playing weight adjuster, does this. Different cartridges weigh different amounts – so to be able to balance them out, the counterweight is adjustable. (The weight of a cartridge is completely different from the playing weight of the stylus: a heavy cartridge does *not* imply a large playing weight.) Some particularly heavy or light cartridges cannot be balanced out on some arms. In some cases, different counterweights can be substituted – if in doubt, check the deck manufacturer's specification, which usually gives the range of cartridge weights that can be balanced out.

Bias compensation (anti-skating force). With ordinary pivot arms, the stylus tends to be pulled – biased – against the inner edge of the record groove, which might lead to more distortion and record wear. On most arms, a counteracting force is applied at the pivot end of the arm. The correct amount of bias compensation to apply is not easy to decide. Bias force is caused by the groove dragging the stylus, so it varies with the type of stylus, type of music, and the position of the arm on the record. Indeed, experts cannot decide whether the bias force increases or decreases towards the centre of the record.

But any bias compensation is better than none, as long as it is not excessive.

The bias adjuster is usually calibrated so that it is correctly set when its scale reads the same number as the number of grams set for playing weight. This calibration is not always accurate. When inaccurate, it tends to be on the high side; so, having set the bias by the manufacturer's instructions, it would be sensible to see if reducing the bias in steps of say 30 per cent, 50 per cent and 100 per cent improved or worsened distortion performance. Again, this is easiest with a test record.

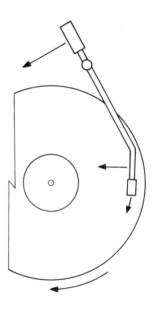

The groove of a record exerts a pull on the stylus. With a conventional angled arm this pull, or side-thrust, is at an angle to the pivot (strictly, to the line joining the stylus point to the pivot), so it tries to pull the stylus inwards, making it press more heavily against the inner groove wall than the outer one. This can lead to distortion and poor stereo effect.

To prevent this, an outward force, called bias compensation, is applied – usually on the other side of the arm pivot, so it pulls in the same direction. Often, the compensation is a simple weight and thread arrangement; springs and magnets are sometimes used instead to supply the necessary force.

Speed accuracy. If the speed is adjustable, it can be set using a stroboscope – decks with adjustable speeds usually come with strobes. But bear in mind the warnings on page 68.

Positioning the deck is an important aspect of setting up. The turntable needs to be level (check with a spirit level), though few decks have adjustable feet to help you with this. The deck also needs to be mounted so that mechanical and acoustic vibrations cannot upset performance – see page 70.

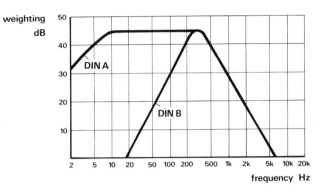

Rumble

Rumble is the background mechanical noise produced mainly by the bearings in the turntable, and picked up by the cartridge. It is aptly named: audible rumble is a low-pitched, growling sort of noise – and like all background noise, it is most obvious during quiet musical passages. In effect, rumble limits the maximum signal to noise ratio of the record-playing system.

Since rumble is mainly a low-frequency noise, it is measured using a weighting which emphasises the low frequencies. Two main weighting curves can be used: *DIN B* results look much better than *DIN A*, and so are the ones you are most likely to find in manufacturers' specifications. For once, though, this is the more useful result to have quoted: it correlates better with the results from subjective tests of audible rumble than does DIN A.

A good result would be 66dB or better. Most records produce more rumble than this themselves, so a deck boasting a better performance than this would rarely sound better in practice. Some experts think that the audibility of rumble is only half the problem, and that even low levels of rumble that are themselves inaudible can affect sound quality in other ways. Still, if records are the limiting factor, it seems that a high level of performance from the deck is unnecessary.

These curves show by how much – and over what range – frequencies are boosted when making rumble tests. DIN A boosts a much wider range of frequencies, and as well as showing up rumble, it also gives information on the low-frequency cartridge/arm resonance – though this is often better evaluated separately.

Wow and flutter

These fluctuations in the pitch of what should be a steady note are described on page 33. Record decks usually produce good results – around 0.1 per cent (measured DIN peak weighted), which should be inaudible. Wow and flutter of much more than about 0.2 per cent is likely to be audible.

Results can depend on the test record used, and any result of less than 0.03 per cent should be viewed with suspicion: it is likely that the test record itself is producing this much wow and flutter.

Ordinary records themselves can produce quite a lot of wow if their centre hole is slightly offset or rather too large. But the wow produced will be at the low frequency of 0.5Hz, which may not be too noticeable.

Wow and flutter is usually measured using a constant tone from a test record. Real records, of course, have varying tones on them, and these place a varying force or drag on the stylus, which tends to slow down then speed up the record. This slowing down and speeding up can produce wow – called dynamic or transient wow. In practice, the turntable motor can often overcome the relatively small drag of the stylus – but the only way of being sure that extra wow is not being created is to carry out a listening test.

Early direct-drive turntables were supposed to suffer particularly from dynamic wow, but there is certainly no reason to think they are poorer than other types these days.

Speed

The absolute speed of a record deck can be measured fairly easily, though any error is rarely likely to be significant (see page 68). Some disc-cleaning devices of the Dust Bug type can slow some turntables down, and might cause the speed to vary between the beginning and the end of a record. This is a test you are very unlikely to see quoted in manufacturers' specifications, and rarely likely to find in reviews – but then it doesn't often present problems.

Occasionally, the start-up time will be quoted in reviews. This is the time it takes the turntable to settle down to the correct speed after switching on. For most people, the result will be irrelevant: most turntables get up to speed in less time than it takes for the pick up to be raised off its rest and placed on the record.

Acoustic and vibration isolation

There is a variety of tests to gain some idea of how good a record deck is at ignoring unwanted acoustic and mechanical vibrations.

Some reviewers simply tap the table that the deck is standing on (with the amplifier volume turned high) and note if this sets up a howl in the loudspeakers. Another possibility is to leave the pick-up resting on a record on a stationary turntable and feed a loudspeaker pointing at the deck with ever-increasing levels of noise. The output from the pick-up (which is all from unwanted vibration, of course) can be analysed and the various resonances at different frequencies can be examined. Some will be more annoying than others.

Sometimes more sophisticated tests are used – but they all have one thing in common: correlating the results with how the deck will actually sound in practice (and in your listening room) is very difficult.

Frequency response

Pick-up frequency responses are rarely completely flat – common problems are a fall-off at high frequencies (say over about 10kHz), or a dip in the mid frequencies, often followed by a peak between 10kHz and 20kHz.

It is not easy to tell, from a printed frequency response graph (whether in a review or from a manufacturer's specification) how a given cartridge will sound in your hi-fi set up. Not only will the loudspeakers affect the response, but so too will the amplifier input loading, if the cartridge is sensitive to this. Very roughly, though:

● a treble peak may give a thin, bright or harsh sound, particularly if it is positioned as low down as about 10kHz; from about 18kHz upwards, a peak should have little effect on the sound (except, perhaps, to increase distortion). The effect of a peak is often exaggerated if there is a dip just before it. On the other hand, with a generally slightly uneven response, a peak might not be so obtrusive as it would if the response around it were flat

● a treble roll-off will make the sound appear dull. Again, the lower the frequency this starts at, and the flatter the rest of the response, the more noticeable this effect will probably be

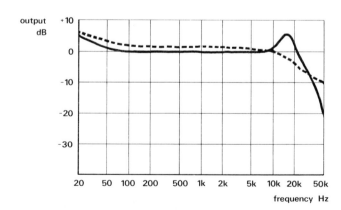

Cartridge frequency response: the solid line indicates a treble peak, the dotted one a treble roll-off

- a dip in mid frequencies may make for a dull sound, or may obscure detail, or may affect the balance of instruments – giving some less prominence than they ought to have.

Ideally, a perfectly flat frequency response should be best – assuming you have loudspeakers with a flat frequency response too. But in practice, you have to choose between one type of fault and another, and to choose a cartridge that complements the response of your loudspeakers.

Separation

Pick-up cartridges are not as good as some other components at keeping the information in the two channels separate. However, most cartridges manage a mid-frequency separation of 25dB or more, which is fine. Even as low as 20dB is usually adequate. Separation usually gets worse at low and, particularly, at high frequencies – one of the effects of the high-frequency stylus resonance. Be careful, therefore, of specifications that quote only the separation at 1kHz: they may be hiding particularly poor figures at higher frequencies. On the other hand, don't look for huge separations at 10kHz or more: not only will you not find them, but they aren't necessary, as high frequencies carry little stereo information. *Which?* tests look at the range 500Hz to 6.3kHz; a separation that doesn't fall below 20dB over this range should be sufficient, though some experts argue that 35dB or more, and over a wider frequency range, is necessary for the finest stereo effect, and for detail in the music to be heard clearly.

Distortion

Compared with some other pieces of equipment, cartridge distortions can appear fearfully high – though in recent years, they have been getting better, and the type of distortion they produce is not so obvious audibly.

Distortion figures seem particularly difficult to compare from review to review, because there are more than a few methods of measurement – depending mainly on the test record that the reviewer uses.

Trackability

Pick-ups cannot trace very high recorded levels accurately – so it is useful to know the highest level that a pick-up *can* trace or track securely, to ensure that it is capable of handling most commercial records.

A test for trackability usually consists of playing a test record with increasingly-highly recorded levels on it, and noting the level at which distortion increases rapidly. Increasing the tracking weight and setting the bias correctly will often help a cartridge cope with higher levels. Indeed, a trackability test is often used to decide what are the best tracking weight and bias level for a particular cartridge.

Some reviewers (and manufacturers) make a lot of fuss about cartridge trackability. But it is not absolutely necessary for a cartridge to be able to track the highest levels on some test records: they don't correspond to the levels usually found on real records. (On the other hand, if your system passes this test, and you still hear distortion on real records, you're probably right to suspect the record itself.) And tracking ability seems to have little correlation with overall sound quality: some cheap and not excellent (though adequate at the price) cartridges show near-perfect tracking ability; most moving-coil cartridges are relatively poor trackers, but have a sound quality that is at least good.

Compliance

It is quite important to know the cartridge compliance because this determines the sort of arm that the cartridge will best partner – see page 70. Unfortunately, compliance is not easy to measure. This seems strange, given the definition of compliance – how easy it is to deflect the stylus by a given amount. This *static* compliance is, in fact, relatively easy to cope with, but it gives no idea of how compliant the stylus is when being moved around by a record groove – *dynamic* compliance.

Dynamic compliance can really be estimated only by examining the behaviour of a cartridge (for example, its resonance frequency) when playing a test record in an arm of known effective mass. Unfortunately, calculating an arm's effective mass is also difficult, and, in turn is often best done by examining *its* behaviour with a cartridge of known compliance.

All this obviously makes specifications and reviews difficult to compare: if you are concerned to match cartridges and arms accurately, our best advice is to use reviewers' stated recommendations as a guide, rather than try to compare test results across several different sources.

Record and stylus care

All records are at least slightly noisy when new – and they get noisier because they get dusty and dirty. It's easy for the smallest particle of dust to get trapped in the disc's groove: when the disc stylus passes over the speck of dust, it can produce a clicking or popping sound through the loudspeakers. Dirt that is ground into the walls of the disc groove can damage the groove, also making the record sound noisy.

To make matters worse, discs are excellent insulators, and readily become electrostatically charged. The disc can then attract small, light particles – dust in particular – which will stick to it. Mostly, it is the attracted dust which then causes the noise – though, in bad cases, the disc itself might have an electrostatic discharge, producing more pops and clicks. Particularly in centrally-heated houses, where the air is usually very dry, it is quite easy for records to become charged; pulling them out of the disc sleeve's plastic inner liner can be enough.

The gadgets. There is a wide range of gadgets to help you keep dirt and dust off your discs, or to prevent them being worn unduly.

The simplest device is some sort of **pad**, **cloth**, or **brush**, which you can wipe over a disc before playing it. Some are designed to be used dry; others with a small amount of fluid of some sort. Fluids are usually claimed to help dispel static rather than to be cleaners in their own right. *Which?* tests show that cleaners used damp tended to make matters *worse* rather than better.

Cleaning arms look and behave like a pick-up arm, but with some sort of a brush or roller in place of a pick-up cartridge; they are used while the disc is playing. The simplest types, used dry, worked fairly well in *Which?* tests – but the more complicated devices couldn't justify their higher costs.

It is usually easy to tell if you have bad static problems: discs will crackle as you lift them from the turntable, and plastic sleeves will cling to them. Dry pads and arms are not a solution; they may even increase static, and the more complicated pads and arms with antistatic devices seem little help either. But if you are plagued with static, you should find that an **electrostatic gun** – you shoot it at the disc before playing – in conjunction with a cleaning arm will help. A more recent development is a **liquid** which you can spray on your new discs and which is claimed to make them permanently static-free – too new for *Which?* to have included in its last tests.

There are gadgets for **restoring** or cleaning very dirty, neglected discs. Some can be quite brutal, and may leave sticky deposits in the grooves – as with antistatic fluids, a bad idea. Some record shops operate a **cleaning service**, using a device that is supposed to suck up all the fluid it uses to wash the disc with.

You can buy **coatings** which claim to preserve records. In fact, groove wear is not often a problem. Unless you have very acute ears, you are not likely to notice much degradation in sound quality over the first fifty or so playings. The coating *Which?* tested gave very variable results: at its worst it gave very much poorer results than doing nothing to the disc.

Record husbandry. Modern gramophone records don't break easily like old 78s. But they're fairly delicate things, with grooves no more than about one-thousandth of an inch deep. Saving these from damage needs a bit of care.
- **Store discs upright.** Stored flat or at an angle, it is easier for them to become warped. Ideally, they should be lightly squeezed together. You can buy spring devices to do this; alternatively, you could use vertical dividers every couple of dozen records or so.
- **Keep records away from heat.** Don't store them near fires or radiators; don't leave them on top of the amplifier.
- **Don't handle disc surfaces.** Grease and moisture from your hands will trap dust and dirt in the grooves. Hold by the centre label and edge only.

● **Store discs in their sleeves.** Use both the inner liner and the outer sleeve, and make sure the two openings don't coincide; this makes it more difficult for dust to enter. (With most sleeves the opening is at the side; having the opening of the inner liner at the top may prevent the drop that makes nonsense of all your husbandry.)

Stylus care. Always put the stylus on the record gently. By the time you can hear your stylus is worn, it's probably already damaged your discs. A diamond stylus should be checked about every 500 sides (and if you ever clout it). You really need a hi-fi dealer equipped with a microscope to do this. Brush loose dust off a stylus with a small, soft brush. Do this gently, always from the back of the stylus towards the front. Harder deposits of dirt on the stylus may need loosening with a drop of solvent first; try isopropyl alcohol or, at a pinch, vodka.

Chapter 7 Radio tuners and FM aerials

For many people, a good radio tuner forms an essential part of their hi-fi set up. Often, it will be part of a tuner/amplifier (sometimes called a receiver). The information in this chapter is equally relevant to tuners and to the tuner part of tuner/amplifiers.

An encouraging start for this chapter: latest *Which?* tests show that there is little difference in performance between many tuner/amplifiers; all are good. So most people, having read the section on features, should have no difficulty choosing a tuner or tuner/amplifier that will be perfectly satisfactory. But if reception is difficult where you live (or you want to listen to particularly distant transmissions), you may need to think harder, with the help of the *Test* sections starting on page 85. And you will need to make sure your aerial set-up is good – see pages 91 to 94.

The chart opposite shows the range of radio programmes in this country, transmitted on many different frequencies (or wavelengths). Frequency is the preferred term these day: you might find medium wave called MF instead of MW, and LW called LF. The M in both cases stands for medium; but the L in LW stands for long, in LF, for low.

The high frequencies of broadcasting transmissions have to *carry* the audio signals (which vary between 20Hz and 20kHz) in some way. There are various methods of *modulating a carrier wave* in sympathy with the audio signals. For broadcasts in the UK, two methods are used: programmes on LW and MW are transmitted by *amplitude modulation*, (AM) and those on VHF (very high frequencies) are transmitted by *frequency modulation* (FM).

FM has had a chequered history. In the 1920s it was thought that FM would have great advantages over AM. In particular, it should take up less frequency space (even in those days, the MW band was getting crowded)

and cause less interference. Later work showed that FM transmissions theoretically required an infinite amount of frequency space, and so would cause *worse* interference.

But the work of America's Major Armstrong in the late 1930s, and the opening of the first FM radio station in New York, showed that FM was possible in practice, and could give advantages over AM – less susceptibility to hiss, fading and interference. But FM broadcasts do use up more frequency space than AM transmissions, so FM cannot be used as the transmission method on MW or LW. The VHF band does give plenty of room for quite a few FM broadcasts; the transmissions take advantage of this, and broadcast a much wider audio frequency range than they are able to do in practice on AM – giving music of hi-fi quality. This is probably one of the main benefits of FM, even though it's only a by-product.

Stereo broadcasts

The agreed system of stereo broadcasting involves transmitting a mono signal, plus additional information which stereo receivers use to extract a stereo signal. This extra information is simply ignored by mono receivers which makes the system *compatible*.

The programme that is to be transmitted is split into two signals: left plus right (mono information) and left minus right (stereo information).

The stereo information then has to be coded: this is done by amplitude-modulating it onto a 38kHz *sub-carrier* frequency. To allow a receiver to de-modulate this coded signal, it has to re-form the sub-carrier exactly as it was transmitted: to help it do this, the transmitted signal also contains a 19kHz *pilot tone* which is synchronized with the 38kHz sub-carrier. This coded signal is then combined with the mono signal; the whole lot is frequency-modulated, and then transmitted.

Although it would be possible to transmit a system of mono-compatible stereo broadcasts using AM, it would mean using up more frequency space for each transmission. On the (already much wider) frequency space each FM broadcast takes up, there is room to fit in the extra stereo information without using much more space.

What wavebands?

All hi-fi tuners receive the VHF waveband – the only waveband in the UK that transmits in stereo, and that is capable of giving hi-fi reproduction. At present, VHF programmes are broadcast over the range 88MHz to 98MHz; over the next few years, the police, and other services, who currently use the range up to 108MHz will move to other parts, and these frequencies will be available (as they already are in many other countries) for entertainment broadcasts. Most tuners can already receive broadcasts up to 108MHz.

The BBC duplicates many of its national and local radio programmes on both VHF and on MW or LW. But not all: Open University and Schools broadcasts currently take over Radio 3 or 4 VHF for parts of the day, for example. This might seem a strange decision: rarely do these broadcasts need the sound quality that VHF can provide, while the programmes they displace (particularly Radio 3) often do. The rationale is that not all listeners have VHF radios, so it is better to put the most popular programmes on the channels that most people use.

Still, it means that if you want your tuner to give you full coverage, it must be equipped with medium and long wave (Radio 4 is on long wave) as well as VHF. Most tuners have MW, but not all have LW. If you find this restricts your choice of possible sets, consider getting a portable radio for AM duties, and widen your choice of tuners to include ones without LW or MW. Since the sound quality on AM is restricted anyway, a portable radio should be adequate – though perhaps irksome if you want to do a lot of recording from AM.

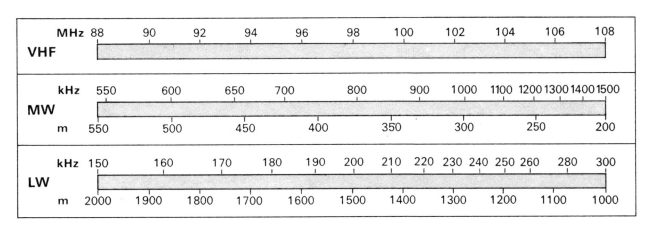

A typical tuning scale, showing the LW and MW bands marked in both metres (m) and kilohertz (kHz), and the VHF band marked in megahertz (MHz). Note that the *frequency* and *wavelength* are related, as they are for all types of waves. Multiplying one by the other gives *velocity*. So, the equation to use for translating from one to the other is (for radio waves in air), frequency (in Hz) times wavelength (in metres) equals 300,000,000 (or 300,000 if frequency is in KHz).

The *Radio Times* contains the BBC frequencies for the area it is published in. The commercial radio equivalent is the monthly *Radio Guide*.

FEATURES ○|○|○|○|○|

There are relatively few knobs and switches on a tuner. Indeed, when it was fashionable for manufacturers to provide matching amplifiers and tuners, the tuner fascia looked very bare. That is slowly changing now; not only because the trend is to miniaturisation, but also because manufacturers are discovering the new miracle ingredient – the micro-processor, which allows hitherto undreamt-of features to be added.

Tuning

Most of the features on tuners are connected with tuning itself: knobs and switches to let you select the station you want accurately – and meters, dials and lights to tell you how accurately you have selected it. This can be carried to such a pitch that one manufacturer actually claimed that you can 'trust them even more than you can your own ears'.

Tuning knobs are found on nearly all tuners, some with conventional mechanical tuning action, some with the latest electronic tuning. It is usually connected to a fly-wheel device, so that when the knob is twisted, it will keep on spinning smoothly. On a conventional tuner, this moves the tuning pointer up and down the scale, and alters the tuning capacitor inside, directly. On electronic tuning, the tuning knob alters a variable resistor, which controls the tuning. There may be a conventional tuning pointer, or an electronic digital readout.

Most people prefer a tuning knob which spins freely, and with a feeling of solidity about it – but there is no particular virtue in this, unless you subscribe to the view that it will be indicative of the quality of the rest of the set (and there is no reason why this should be so). As long as stations can be tuned in accurately, the tuning system is doing its job.

Tuning scales should be easy to read, but the display if lit should not be too bright. There is no need for great precision in the marking on the scale – few tuners have their pointers perfectly aligned against the right frequency, and in any case you are more likely to use your ears, or a tuning meter device, to tell you when you are exactly on station. Much of this applies to digital readouts, too – here, because the actual frequency is flashed up before your eyes, there is an even greater tendency to demand unnecessary precision in its reading. (Precision in the tuning itself is a different matter – see below.)

Preset and automatic tuning. Preset tuning means a number of stations can be tuned in and then any one selected by pressing the appropriate preset button. No self-respecting television would be seen dead without half a dozen or more tuning presets, but few radio tuners have them, and almost no Japanese ones – partly because their biggest market is America, where there are so many stations that preset tuning becomes impossible. Check the number of presets available, and see which can be used on FM, or only on AM (sometimes a preset can be swapped between the two). Electronic tuning makes it easier to provide preset facilities – but if the tuning frequencies are stored electronically, the set will probably need a back-up battery to prevent the information being lost if the power should fail, or if the set is unplugged.

Electronic tuning will also make it easy for tuners to do tricks – such as sweeping the waveband automatically, looking for stations; letting you dial station frequencies directly on a calculator-style keyboard and so on. These sorts of facilities are likely to cost a lot of money – at least in the early days. And most people will probably find they can live without them. (Think first how often *you* change between more than, say, four regular FM stations, and how often you want to search the whole of the frequency band.) An exception might be the *DX-er* – someone who treats looking for different and far-away stations as a hobby. Sophisticated electronic searching devices might then have a job to do (though a true DX-er will probably consider them in the same light as a mountaineer would a portable lift).

Preset tuning is a boon: with it you can easily and quickly change between the main broadcasts that you listen to. Automatic tuning features are less necessary,

though – unless you spend a lot of your time searching for different stations.

Tuning indicators. Most, if not all, tuners provide a *signal strength meter*. These are of little use: most of them read their maximum at fairly low signal levels, so they cannot be used to tell if the aerial is as large as it needs to be, if it is accurately aligned, or if the set is very accurately tuned. In any case, on FM the largest signal does not necessarily give the best audio quality.

Better is a *centre-zero meter* (FM only): when the meter's needle is dead-centre, the set should be on tune. Many tuners will have both types of meter. To save meters, some sets double up: a single meter does as a centre-zero type on FM and signal-strength type on AM; or on tuner/amplifiers, the tuning meters may also function as power meters.

Some tuners have two or three *lights* instead of meters. The set is tuned when the lights are equally bright (if there are two) or when the centre one is at its brightest (if there are three). So the lights act as a centre-zero meter.

Rare are meters which measure, in effect, *signal to noise ratio* – useful if you find noise more oppressive than distortion (though again, your ears should tell you).

More complicated indicators can be useful occasionally – for example, when setting-up an aerial – but are probably not worth going out of your way for. Most common (though still fairly rare) is a *multipath* distortion meter – see page 92 for details of multipath distortion. One tuner includes an oscilloscope for displaying tuning information – expensive, and probably more for pleasure than business. In case you can't hear how much multipath distortion there is, another tuner has a switch that cuts out the music, and lets you listen just to the distortion.

Tuning indicators can be useful for accurate tuning: even slight inaccuracies on FM rapidly increase the distortion level. Great sophistication is unnecessary, though: some form of centre-zero meter or light is really all that is needed. (In any case, you could argue that the best tuning indicator is your ear, and you tune until the sound appears least distorted.)

Automatic frequency control (AFC) locks the tuner onto the selected station. Few sets suffer from tuning drift these days, but AFC can still be useful to correct for poor tuning. With some tuners, the AFC action can be rather fierce: if you want to tune into a weak station near a strong one, the AFC might prevent you – altering the tuning so as to receive the strong station. There is no problem if the AFC can be switched off.

Muting. A loud shooshing noise can be heard between FM stations. A muting circuit simply switches off the output between stations to get rid of this irritation. The trouble is that it might mistake a very low-level station for noise, and mute that as well – no problem if muting is switchable. Often the muting switch is combined with a stereo/mono switch: the muting is always *on* in stereo, and *off* in mono. The theory is that if you want to receive weak stations, you will be happy to do so in mono, which will reduce the hiss; if you want comfortable stereo listening, then you need inter-station muting. A reasonable theory, but you may prefer a tuner where the choice is left to you.

Switchable bandwidth. The wider a tuner's bandwidth, the less distortion there will be – but the poorer the selectivity. (Selectivity is a measure of how well a tuner can do its main job – picking out just the station that you want to listen to. See page 87 for details.) A switchable bandwidth should give you the best of both worlds. For strong local stations, where there is little chance that stations near by in frequency will be strong enough to interfere, the wide setting can be used to get the very least distortion. For picking up distant weak stations, the narrow setting will help ensure that stronger stations do not interfere.

For most people, though, a tuner without switchable bandwidth will give perfectly adequate levels of distortion and selectivity. So this is not a very important feature unless you want to listen to both very distant and very local stations and have a critical ear for distortion. Switchable bandwidth may still not help if the two positions do not give usefully-different levels of distortion and selectivity: check this in a home trial if you can.

Stereo

All tuners can be switched to mono, so that broadcasts, particularly weak ones, will sound less hissy. **Variable separation** provides a variable amount of stereo – so you can choose your best compromise between stereo effect and noise. A better approach to the problem, and more useful, is a **high-blend switch**. Only the higher frequencies are reduced to mono: this can reduce hiss (which is mainly noticeable at high frequencies) quite a lot, but hardly affects the stereo image (which is mainly *not* noticeable at high frequencies). Many high-blend switches are labelled **MPX** (for multiplex).

Stereo indicator lights show the presence of a stereo programme. There are two philosophies about this. The light might switch off when the tuner is switched to mono – so there is no confusion over whether you are listening in stereo. Or it might stay on when the set is switched to mono (provided a stereo signal is being broadcast, of course) – perhaps more confusing, but slightly more useful, because it shows that you could listen in stereo if you wanted to.

Inputs, outputs, switches

The only input is an aerial socket or two. For FM, there are two main types, described by their input impedance. For proper outdoor or loft aerials, which will be connected to the tuner by co-axial cable (popularly known as *coax*), use the 75-ohm terminals. With a bit of luck, the tuner will be equipped with a proper coax socket – like the one on a TV. More likely, there will be some form of screw terminal arrangement, which can be a bit fiddly to connect up; as it is a job you won't be doing very often, it is not much of a burden. The 300-ohm socket is used with continental aerial leads, and the ribbon aerial that is often supplied with a tuner. Some Continental sets have only 300-ohm terminals: connecting a 75-ohm terminal to these effectively makes the tuner half as sensitive, which might make already-weak signals sound rather hissier, but will not affect strong signals very much. A *balun transformer* connected to the input terminals changes the 300-ohm impedance to 75ohms, and restores the effective sensitivity. A balun will cost a pound or two – but is almost certainly cheaper than buying a more sensitive tuner, and probably cheaper than buying a more powerful aerial.

It is rare for tuners to have AM aerial sockets. Instead, there is a built-in **ferrite aerial** – usually positioned on the back of the set. These are very directional: for best reception (and AM reception is often pretty poor even at the best of times) it needs to be rotatable *horizontally*. Many simply swing out from the back of the set, which is of little use unless your set just happens to be pointing in the right direction.

FM aerials are directional, too. For best results they have to be directed at the transmitter: difficult if two transmitters are not in the same direction. In these cases, an electrically-driven *aerial rotator* is an expensive but useful accessory. Even more luxurious is a tuner that *automatically rotates* the aerial round to the correct position, depending on the frequency it is tuned to. One or two models are available.

Tuner output levels are rarely likely to cause any distress to amplifiers they are plugged into – but for the odd matching problem a *variable output level* could be useful.

An *air-check* sounds a strange device; *recording level check* would be a better term. When switched on, it provides a tone at 75kHz deviation (ie full modulation), so that you can ensure your recording level controls are set to get the maximum amount of signal on to the tape (and hence the best signal to noise ratio for the least amount of distortion). But, as explained in the chapter on tape decks, the maximum amount of steady tone a deck can cope with is not much of a guide to what it can do on real music with transients. Not all radio stations consistently peak near full deviation, anyway. So, only slightly worth having.

There is no need in the UK at the present for the *switchable de-emphasis* found on a few tuners – as the box opposite explains.

TESTS ☑ ☐ ☐ ☐ ☐ ☐

Measurements on tuners fall into two (somewhat connected) categories. *Radio frequency (rf)* or front-end measurements concern the ability of the tuner to pick up the wanted signal and sharply reject unwanted signals, interference and so on. *Audio* measurements concern the audio quality of the output signal – its frequency response, the amount of distortion and noise present (though these factors are largely determined by the front end, in fact) and the general accuracy of the sound.

Sensitivity

This much-quoted measurement deals with how well the tuner picks up weak broadcast signals. There are (as usual) different measurement methods, giving different, and not always comparable results. But the result is always given in the same form – as the amount of signal needed (usually quoted in μV) to give a certain signal to noise ratio at the output. So the lower the figure, the better. To compare different figures properly, you would need to know two things.

● The amount of modulation or deviation in the broadcast signal – since this affects the measured signal to noise ratio. The best result (ie the lowest figure for sensitivity) comes with using the highest deviation allowable – 75kHz. So if the deviation level isn't given (and it rarely will be), it is probably best to assume 75kHz. Other levels sometimes used, 40kHz and 22.5kHz, obviously result in poorer sensitivities (so long as the noise is measured in the same way), though how much poorer will vary from tuner to tuner. Very roughly, however, sensitivity figures using 40kHz deviation will be about half as big again (remember, the lower the figure, the better) as those at 75kHz.

● How the noise is measured. The full battery of different noise measuring methods – see Chapter 3 – is used in different Standards. As with other measurements of noise, CCIR weighting (either ref 1kHZ or ref 2kHz) is probably the most sensible to use. CCIR ref 2kHz gives sensitivity figures roughly 25 per cent lower than using CCIR ref 1kHz.

If you see a figure for just *sensitivity*, with no more explanation, the result is probably that for the **ultimate sensitivity**. The signal to noise ratio used for this sensitivity measurement is about 30dB – very hissy. Not only are

---

Pre-emphasis and de-emphasis

To improve the signal to noise ratio of a broadcast (ie to make the noise less apparent), either the signal could be increased, or the noise could be reduced. The amount of noise depends partly on the electronic circuitry and components; there is a limit to how much this can be reduced. The amount of signal is limited because the loudest sounds must not cause more than 100 per cent modulation.

Noise is most apparent at high frequencies where, in general, the sounds are not very loud. So broadcast signals can have the high-frequency part of their signal artificially boosted on transmission (by 10dB at 10kHz) without causing over-modulation. This is called pre-emphasis, and helps to increase the signal to noise ratio by about 5dB overall. the tuner then has to reduce the high frequencies by the same amount (de-emphasis) to get the frequency response flat again.

Electronic engineers often express such boosts or cuts in terms of a *time-constant* – this is a way of specifying the relative values of the resistors and capacitors which go to make up the boosting circuits. In the UK and Europe, the time constant is 50μs (microsecond – one millionth of a second). Because this is standard, *switchable de-emphasis* is not something you need look for on a tuner. (In America, it's 75μs: Americans boost the high frequencies more than we do.)

CHAPTER 7

TUNER CHARACTERISTICS

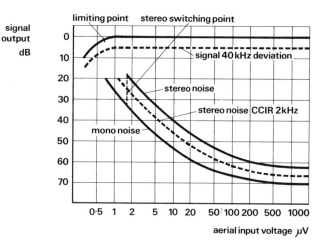

Along the bottom axis is plotted the aerial input voltage – ie the amount of signal in microvolts at the aerial input. Up the side is plotted the strength of the audio output in dB. The solid top curve shows how the audio output varies with different signal strengths. At very low signal strengths, the output increases as the signal increases, but it very soon settles down to a maximum value. This is a consequence of *limiting* the signal to stop AM-type interference. It might be thought to be a nuisance – after all, if the output was allowed to rise, then in strong signal areas a large output would be obtained which would need less amplification. But the advantages of less interference, and an output which does not vary in strength with variations in aerial signal, easily outweighs this.

The bottom two solid curves show what happens to the amount of *noise* in the output signal: it drops as the input voltage increases, so even though the output signal does not increase, the signal to noise ratio gets better as the aerial signal gets stronger. The noise drops more rapidly at low signal strengths than at high ones: by the time the input voltage is around 1mV (ie 1000µV) or so, the noise has reached its minimum value – so there is nothing to be gained from trying for larger input signals, for example, by using a more powerful aerial.

Note that the noise in stereo is greater than it is for the same signal strength in mono. At high signal strengths, though, the difference is usually fairly small, and may not be noticeable.

From the graph, all sorts of figures may be read off. For example, the *signal to noise ratio (S/N)* for any particular signal strength is merely the distance in dB between the signal level at that point, and the noise level – using the stereo or mono curve, depending on which signal to noise ratio you want. For example, at 1mV, the stereo signal to noise ratio is 63dB, and at 1µV, the mono ratio is 25dB. Similarly, the *sensitivity* for any given signal to noise ratio is the input signal corresponding to that signal to noise ratio. For example, the ultimate sensitivity (for 30 dB S/N mono) is at 1.5µV, and for 50dB stereo S/N is about 30µV – very good.

The graph can tell us more. For example, it is useful if the *limiting point* (the point at which the output is 3dB below its maximum level) occurs at an input voltage lower than the ultimate sensitivity, as it does here – then all signals that can be heard will be received without fading.

The graph also shows the *stereo switching point* – where the tuner will automatically switch from mono to stereo (assuming that a stereo signal is being broadcast, and that the tuner is switched to stereo). It is not sensible for this to happen at too low an S/N ratio, as bad hiss will probably mar any enjoyment of the stereo effect – though you could always switch to mono manually if you want. On the other hand, if the stereo switching point is set at too high an S/N ratio level, you don't get the option of listening to weak stations in stereo. Most tuners have the switching point at around 20dB, which is probably as low as necessary.

The broken curves show the effect on the results of using some different measurement methods. Lowering the signal from 75kHz deviation to 40kHz gives an output a constant 5½dB lower, so all signal to noise results look 5½dB worse. Any sensitivity measurements will give poorer-looking results, too – but how much poorer depends on how quickly the noise drops. In this example, the sensitivity for 50dB S/N stereo rises from 30µV to 70µV, but if the noise curve were steeper, the increase would be smaller. Changing from CCIR ref 1kHz to CCIR ref 2kHz changes the noise (and hence signal to noise ratios) by a fixed 6dB: again, the effect on the sensitivity results depends on the slope of the noise curve.

RADIO TUNERS AND FM AERIALS

you unlikely to want to listen much at this sort of noise level, but the results are so good that they are largely academic – around 1μV or so.

A slightly different measurement is that of **least usable sensitivity**. Instead of using *noise* to decide how poor the output can be, this method uses noise plus *distortion* (by measuring the output with a distortion factor meter). The results are just as academic, however.

A slightly more useful figure is the sensitivity for a signal to noise ratio of about 50dB in stereo – for most people, a just tolerable noise level, and so a realistic sensitivity level. Reasonable hi-fi tuners will turn in figures of between 200μV and 50μV on this test – depending on the tuner, and the method of test. As with many technical tests, small differences between tuners are not important.

Many people will find even this sensitivity figure of little practical importance. Broadcasting authorities try to get a signal level of *1000μV* over the area that each transmitter is designed to serve. So, anything poorer than about 300 or 400μV will matter only to someone who *wants* to receive weak signals – because, for example, they want to receive a particularly good but far away local radio station, or a continental transmission; or someone who *has* to receive weak signals – perhaps because they live in a basement flat and cannot put up a good aerial, or because they have no transmitter station near.

Selectivity

Modulating a carrier creates a complicated signal which contains side frequencies above and below the main carrier frequency. So a tuner set to 91.3MHz, say, needs to pick up all frequencies from about 91.2MHz to 91.4MHz – and, so that it does not pick up any neighbouring broadcasts, it should sharply reject all other frequencies.

Selectivity describes how well a tuner does this rejecting, by measuring how large a near-by signal has to be before it interferes noticeably with the wanted signal. Of course, selectivity can be over-sharp if, in rejecting unwanted frequencies, the tuner rejects part of the frequency range of the wanted signal. In this case, the distortion performance of the tuner would be poor.

The frequency band is thought of as being split up into *channels*, each 200kHz wide. The channel next to the one containing the wanted signal (centred 200kHz above or below it) is called the *adjacent channel*; the next-but-one channel (centred 400kHz above or below the wanted signal) is called the *alternate channel*. Selectivity test methods measure alternate or (more rarely, since the figures are much poorer) adjacent channel selectivity.

Selectivity in practice

In practice, the frequency band is not split up into neat channels; there are far too many stations competing for space. Most stations fall at 0.1MHz (ie 100kHz) intervals, but each of *these* frequencies is allocated to more

---

Pd, emf and power

The results for sensitivity will sometimes include the letters *pd* or *emf* after the figure. Sensitivities quoted as so many μV (pd) give results half as big (and therefore twice as good) as μV (emf), so is the one likely to be used in specifications.

The sensitivity in μV (pd) depends also on the resistance of the aerial inputs. So the sensitivity of a tuner at its 300-ohm input will always be twice its sensitivity at its 75-ohm input, if the tuner's input circuits are properly designed. This is a result of the measurement method, and does not mean that the tuner performs worse on 300ohm than 75ohm.

To get over this confusion, some standards (including IHF) quote sensitivity in terms of the generators' available output power, which is irrespective of the aerial input resistance. This power is usually given as a dB ratio, where 0dB is one *femtowatt* – one thousand million millionth of a watt. The symbol used is dBf, or dB (fw).

than one transmitter, and a few stations are at closer intervals.

So why isn't there intolerable interference? Because the network of transmitter stations all over the country is carefully planned so that those which are close together in frequency are usually widely separated geographically. Within each area, transmissions on frequencies close to a wanted frequency are very weak and so will be ignored, and all the wanted close stations, whose transmissions are very strong, are widely separated in frequency.

This theory works well, but the practice is not perfect. For example, in London these are the local transmissions that a tuner would be expected to cope with.

R 1/2	89.1 MHz	Radio London	94.9 MHz
R 3	91.3 MHz	Capital Radio	95.8 MHz
R 4	93.5 MHz	LBC	97.3 MHz

The closest pair is 900kHz apart – well separated, even for a tuner with poor selectivity.

But if you lived between London and Ipswich, say, and had a set with good sensitivity, you *might* pick up, at equal strength, LBC at 97.3MHz and Radio Orwell at 97.1MHz – only 200kHz apart. So you would need a set with good adjacent channel selectivity, to prevent them from interfering with each other. (Alternatively, if LBC and Orwell are in opposite directions from your house – as they probably would be – a good aerial may give all the selectivity that is needed – see page 91 onwards).

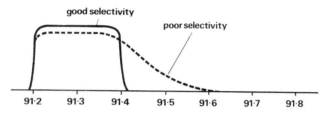

The selectivity of two tuners. The tinted boxes show a station centred at 91.3MHz (the one the tuners are tuned to), and others at 91.5MHz and 91.7MHz (unwanted signals in the adjacent and alternate channels). The solid line shows a good tuner's response: it envelopes all the bandwidth of the station at 91.3MHz, but hardly any of the adjacent channel station. The dotted line shows a poor tuner: it envelopes much of the adjacent channel station, and also responds a little to the alternate channel signal.

Which? selectivity tests are based on the DIN method. A signal generator provides a wanted signal at a frequency of 95MHz; this station carries modulation, as a real station would. The unwanted signal is swept in frequency from 94.4MHz to 95.6MHz, and the amount of interference it causes the signal at different frequencies is measured. As the unwanted signal gets nearer and nearer the wanted one in frequency, the amount of interference increases. In particular, the frequency which causes 30dB of interference to the wanted signal (which is the same as saying that it degrades the signal to noise ratio of the wanted signal by 30dB) is noted down. This sweep is repeated at lots of different input levels – giving a list of figures of the frequency for a particular input level (relative to the wanted signal level) producing 30dB interference. The level corresponding to frequencies 200kHz on either side of the wanted frequency gives the adjacent channel selectivity results; those corresponding to frequencies 400kHz on either side give the alternate channel results.

Good selectivity, like sensitivity, is likely to be necessary only if you want to pick up broadcasts that are not really in your transmission area (and sometimes not even then). Most people won't need good selectivity, and can look for a tuner giving say 45dB or more alternate channel, and 5dB or more adjacent channel selectivity (measured using DIN or IHF methods). There should be few tuners giving results poorer than these, anyway.

Capture ratio

A basic feature of FM is its ability to reject weaker unwanted signals on the *same* frequency as a wanted signal – the FM capture effect. The unwanted signal needs to be only slightly weaker than the wanted one for the tuner to be able to reject it completely – how much weaker is a measurement of capture ratio. It is occasionally called *co-channel selectivity*. In many ways, it is similar to adjacent and alternate selectivity. The measurement is much the same, too. Wanted and unwanted signals of the same frequency are generated, and the

RADIO TUNERS AND FM AERIALS

level of the unwanted one is increased until it causes 30dB interference on the wanted signal. In effect, the difference between the two of them is then the capture ratio. Test methods can be complicated, and express this difference in many ways. Often, the capture ratio is expressed as half the difference between the wanted and unwanted signals; 3dB or *less* (the unwanted signal is smaller, this time, than the wanted signal) is reasonable.

You might think that only people that can receive transmissions on the same frequency would need a good capture ratio – and that there would be precious few of them. That is true – but *multipath* effects also demand good capture ratio for their rejection. (Multipath is explained on page 92: it can occur in areas where the tuner aerial is surrounded by tall buildings or hills.) Even so, few modern tuners have a poor capture ratio.

Interference rejection

Although FM is inherently good at rejecting many forms of interference, excessive amounts can still break through and cause problems.

If the tuner doesn't reject interference derived from the **intermediate frequency** (see box, next column), you may get trouble, if you live near an airport, for example – or you may get whistling noises on stations close to 96.3MHz. Specifications are unlikely to help; listen.

Impulsive interference – especially from near-by traffic – can plague pretty well any tuner. A good aerial is the only answer.

A tuner may pick up **radio frequency intermodulation distortion (RFIM)** – often a burbling sound near the station you're after, even breakthrough into an actual programme. It happens when a weak station suffers interference from several other, stronger signals. A less powerful aerial (or *aerial attenuator*) could stop the interference – but that might be no good if you wanted to receive both strong and weak stations. Figures are rarely quoted in manufacturers' specifications.

Even if you do all the right things, you may still be plagued with interference, even other transmitters breaking through. In some cases, you can get free official advice: ask at a main Post Office for their leaflet, *Good radio and tv reception*.

Intermediate frequencies

Two main jobs of a tuner are detection (recovering the audio signal from the modulated carrier wave) and *tuning* (selecting which of the many broadcasts is to be picked up). Both these jobs are relatively easy, but it does make the design of the rest of a tuner easier if the frequency passed on from the tuning section can be a fixed one. This is the *superhet* principle of design, and the fixed frequency passed on is called the *intermediate frequency* or *IF*: in modern FM sets, the IF is 10.7MHz.

If you are content to receive only broadcasts that are designed for your transmission area, then there is no need to look for particularly good results in the various forms of interference test. Even if you do want to receive weak signals, then some forms of interference can be kept at bay simply by using a set with good sensitivity or (probably better) by using a powerful aerial – perhaps rotatable. But too large an input signal can actually cause some forms of interference – RFIM in particular – so if you want to receive both weak and strong signals, you may need a set good for RFIM performance.

Specifications vary a lot in the types of interference test results that they give: but if you need good performance, look for roughly 80dB or more for any of the types. Better, try to arrange a home trial of the set you are interested in: laboratory measurements cannot predict exactly how a tuner will actually perform where you live. If you want good interference rejection, you will probably need good sensitivity and selectivity, too.

Noise

Signal to noise ratio forms part of the sensitivity measurement – see page 85. The result depends on the aerial input voltage used – normally, a standard 1mV. Many tuners will, in practice, be receiving signals of at least this level, and higher input levels generally bring little improvement in signal to noise ratio.

Unlike interference measurements, where it is difficult to give useful figures, tuner signal to noise ratio is something you can really get to grips with. The best signal to noise ratio that broadcasting authorities can manage (at the moment) is about 69dB (CCIR ref 2kHz). The tuner needs to be a little quieter than this, because the noise of the tuner and the broadcast will add to each other; even so, a result of better than 73dB is all you will need. Unless you are particularly worried by noise, you should find even 66dB to 69dB hiss-free.

Some tuners produce hum: one particularly annoying variety is **modulation hum**, where the hum varies with the frequency content of the programme. The CCIR weighting tends to ignore any hum, and measurements of hum itself are fairly rare. But some specifications and reviews might quote *unweighted* signal to noise ratio as well as weighted. Normally, the unweighted ratio is a few dB better than the weighted – but if there is much hum present, the figure may be poorer.

Frequency response

The audio part of an FM broadcast extends only up to 15kHz. Tuners have to be designed to pass on frequencies up to this level, but then very steeply cut the response, to reject the stereo signal's 19kHz pilot tone. Tuners with poor 19kHz filters will have frequency responses that are already a few dB down at 15kHz. Most people would find such a tuner dull-sounding in comparison to one whose frequency response was still flat at 15kHz. A roll-off at the low-frequency end – below 40Hz, say – is not so important. *Stereo blend* filters and the like should not (but sometimes do) affect frequency response as well as channel separation. Specifications are unlikely to admit to this; check by listening.

Cross-talk (or channel separation) is as important in tuners as amplifiers. And the same points apply: look for a result quoted over a range of frequencies (say 500Hz to 6kHz, or maybe even 10kHz). A minimum separation of 30dB over this range would be very good.

Many tuners used to suffer from a high level of **distortion** in the cross-talk. Cross-talk distortion measurements are rarely if ever quoted, but modern tuners are rather better (as at most things).

It can be useful to know what levels of the two tones associated with the stereo signal – at 19kHz and 38kHz – are left in the audio signal. The 19kHz signal might be audible to the young, and both signals can interact with tape recorder bias frequencies, causing a whistle on tape recordings. Many tuners manage to keep the 19 and 38kHz tones 55dB or more below the level of the signal – sometimes as much as 70dB or more. Anything better than 55dB should be fine; anything lower *might* cause problems, depending on your tape deck, and it would be best to check by making a trial recording.

These tones could also cause intermodulation distortion with the high-frequency part of the audio signal. The distortion frequencies created could be lower than 15kHz, and so could mar reproduction. Distortion tests (better still, listening tests) are the best way to find out.

Distortion

As with amplifiers, the distortion performance of tuners is getting much better, and simple measurements of harmonic distortion (which is all most manufacturers' specifications will tell you) are becoming less and less useful guides to what the tuner will actually sound like.

Aerials for FM

In order to work, all radios need an aerial (or *antenna*, as they say everywhere except Britain) of some sort, designed to pick up the range of broadcast frequencies that the set handles. Even portable transistor radios need an aerial, but because this is small and can be built into the case, it isn't noticeable. Perhaps it is because such aerials are taken for granted that few people pay enough attention to using a proper aerial with a VHF hi-fi radio tuner.

But a good aerial is important for best reception. It effectively increases the sensitivity of the set, and can often be a better solution to poor reception than buying a more sensitive tuner.

Types of aerial

Aerials for VHF are of three main types.

Indoor ribbon aerials are effective only if you live quite close to the transmitters. Otherwise, you will not be getting the best performance the tuner is capable of. This sort of aerial is often supplied with a tuner, and it may do until you are able to buy something better.

Connect it to the 300-ohm aerial terminals on the tuner. If there aren't any, you should really interpose a *balun transformer* if you want to get the best sensitivity possible. Better still, put the couple of pounds or so a balun costs towards a proper aerial.

Outdoor aerials are the proper solution to the problem of picking up signals. It is more or less essential to use an outdoor aerial (or at a pinch a loft aerial – see below) for even reasonably hiss-free listening unless you live quite close to your transmitters. Sadly, *quite close* is difficult to define. It might be 20 miles if the transmitter is powerful and reception conditions are good; if not, it might be less than 4 miles. Aerials of different strengths are available to help you cope with even the most difficult conditions; and a well-chosen and installed aerial will help reduce interference as well as hiss. There are two snags: you may need help to decide which is the best type of aerial for your conditions, and you may need help to erect it.

Connect, probably, to 75-ohm terminals – again using a balun if the tuner doesn't have them. In this case, a balun is probably a cheaper solution than a more powerful aerial.

Indoor loft aerials are a compromise, in efficiency and ease of erection, between a proper outdoor aerial and an indoor ribbon type. There is no need to get a special loft aerial: an outdoor one, correctly chosen, will do just as well. You still need advice on how to select the best type.

Aerial directivity and gain

Aerials are usually *directional* – better at picking up signals coming from some directions than from others. The simplest aerial – a horizontal rod called a **dipole** – gives its greatest output when the signal is coming towards its side, no output when the signal is coming towards the ends, and some output when the signal is coming at an angle to the rod. So, for best results, the aerial must be aligned *broadside on* to a transmitter, *not* pointing towards. (If you do not know what direction the transmitter is in, you can find out by turning the aerial around until reception is at its best, with minimum noise and distortion.)

A simple dipole can be made more effective by adding **parasitic elements** to it. These are not electrically connected to the dipole, but help it capture the broadcast waves. The main parasitic element is the *reflector*: this is placed behind the dipole (ie on the side furthest away from the transmitter) and effectively bounces back the waves on to it.

Other elements – maybe three or four or more, called

directors – can be added in front of the dipole. An aerial like this is called a *Yagi*, after its inventor, and it is still mounted so that the dipole is broadside-on to the transmitter.

For a rod pointing in the right direction, adding elements *increases* the signal fed to the tuner. The amount of this increase, relative to the signal fed by a simple dipole, is called the aerial's *gain*. For example, adding a reflector increases the signal by 1.4 times; the gain is 3dB.

Gain is a good thing if the signal is weak: it can boost the signal more than buying a more sensitive tuner.

Adding elements also reduces the response of the aerial to signals arriving at the back of the aerial. This in turn means that the ratio of the response from the front to the response from the back (the *front to back ratio*) can be high – perhaps as high as 12dB (a simple dipole, has a front to back ratio of 1:1, which is 0dB). Unlike a dipole, there may be some response to signals approaching multi-element aerials from the side – but because the response from the front is even greater, the ratio between front and side pick-up is also greater.

In short, a multi-element array is much more directional than a dipole, and the larger number of elements, the more directional it is.

Directionality is usually a good thing. The aerial can be pointed at a wanted transmission, and unwanted transmissions, noise and interference from all other directions will arrive at the tuner relatively weaker than they would do with a less directional aerial. That makes the tuner's job of selectivity and interference rejection

REJECTING INTERFERENCE

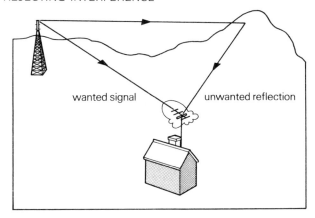

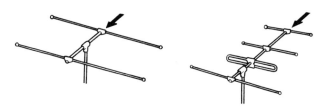

Two VHF aerials. The arrows show the direction that the wanted signal is coming from (ie the arrows point to the *front* of the aerial). **Left**: a dipole aerial with a single reflector and no directors. **Right**: a four-element array – two directors in front of the dipole, the dipole and a reflector. Note that the dipole is bent over on itself. This is because adding elements reduces the impedance of a simple dipole; bending it round to make a *folded dipole* restores the impedance to around 75ohm.

One particularly difficult form of interference is that caused by signals on the same frequency as the one you are trying to pick up – caused by a transmitter working on the same frequency but located in another area or, as in the diagram, by the signal from your transmitter being reflected off tall buildings or hills. These will arrive at the aerial weak, out of phase and slightly later than the main signal – and they can cause an unpleasant form of interference known as *multipath distortion*. The FM system is good at rejecting weaker unwanted signals on the same frequency as a wanted signal – but it can still use some help from a good aerial array.

The answer is to use a very directional aerial, pointed straight at the wanted transmitter. The aerial gives a large output from signals coming from straight in front. This is shown by the *polar diagram* superimposed on the aerial; the distance from the centre of the dipole to the edge of the polar diagram is large. The output due to the unwanted reflection coming in from the side is small; the distance from the centre of the dipole to the edge of the polar diagram is small. The ratio between the strong wanted signal and its reflection will then be very large, even before the two signals reach the tuner front end, and so the job of sorting them out in the tuner will be much easier.

RADIO TUNERS AND FM AERIALS

rather easier. Again, a good aerial may do more for selectivity and interference rejection than changing to a more sophisticated tuner could.

However, if you want to receive transmissions from stations that are not in line with each other, an over-directional aerial might make it more difficult to pull them in – because you cannot point the aerial directly at both simultaneously. One solution is to go for a less directional aerial – though, as the right-hand diagram illustrates this may ensure that all transmissions are equally poorly received; or perhaps fit an aerial rotator (see page 84). You could even fit two separate aerials; if so, you will need to take care in mounting these so that they do not interact with each other.

Which aerial?

It is not easy to decide how big your FM aerial should be. First, it is unlikely that there will be a lot of radio aerials in the vicinity, to give you a clue (as there usually are for TV). In turn, that means the ones that are there may be no good guide to what is needed. Second, reception conditions may change even over a distance of a few yards – so general guidance on what sort of aerial is needed may not be relevant to your particular location.

One way of tackling the problem is to use an **aerial installation firm**. They then have all the bother of working out what size aerial to use, and they face the danger of clambering about on the roof. The snag is that you don't know how accurate the firm's solution will be, and

AVOIDING UNWANTED SIGNALS

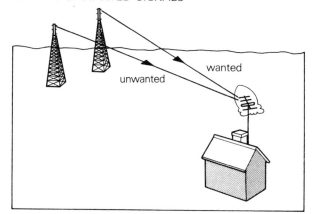

In case you thought aerial installation was easy, the diagram shows an example of when it is *not* a good idea to point an aerial directly at the wanted station. Here, there is a strong unwanted signal in almost the same direction as the wanted transmission. Pointing the aerial directly at the transmitter would mean that the ratio between the wanted and unwanted signals would be quite low. Turning the aerial a little certainly reduces the strength of the wanted signal a little – but, as the polar diagram shows, it reduces the unwanted signal by far more, and that makes the job of the tuner in sorting them out much easier.

PICKING UP DIFFERENT TRANSMITTERS

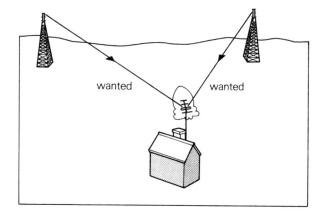

When transmitters you want to pick up are in different directions, pointing your aerial directly at one of them may drastically reduce the signal you receive from the other. It's usually better to share out the misery by pointing the aerial in a direction midway between the two transmitters. Both signals will be reduced but, with any luck, not by too much; you may not even notice any extra hiss. Using an aerial that isn't too directional (a two or three element one) may help too.

If you are a long way from the transmitters, this method will not work, and you'll have to use a directional aerial with a lot of gain and a rotator, or two, with each pointing at a different transmitter.

you have to pay for labour and, perhaps, forego discounts on the price of the aerial.

Another solution is to use an **aerial supply firm**. Some of these give advice on the type of aerial you should have, and may even swap it for another, if it proves unsatisfactory. Try local dealers, or look through hi-fi magazine advertisements. Using the right firm should ensure you get the correct aerial, and you should save some money. But of course, you have to spend time on the job, and you have to take the risk of clambering about on the roof and making sure the aerial is secured properly. You also need at least one helper (with good ears), to listen to the reception and tell you when you have the aerial pointing in the best direction.

The Broadcasting Authorities are worth writing to for help.

BBC Engineering Information Department
BBC Broadcasting House
London W1A 1AA
Engineering Information Service
Independent Broadcasting Authority
Crawley Court
Winchester
Hants SO21 2QA

Both information services can give you general advice on the sort of aerial you need; the BBC also provides maps which show the service areas for particular transmitters. The BBC has also issued a booklet *How to get the best out of BBC stereo radio*. It is free, but send a large, stamped addressed envelope. Most of it is applicable to getting the best out of IBA transmissions, too – or there's IBA's leaflet called *Reception of ILR*.

Chapter 8 Cassette decks and tapes

The first recorder was probably made by a Dane, Valdemar Poulsen, in 1898. He used steel wire, rather than the magnetically-coated plastic tape used today – but the principle was the same: a varying electrical signal flowing in a coil of wire creates a similarly-varying magnetic field. That magnetic field can be transferred to a magnetic tape or wire, if it is pulled past the coil (recording). Then the magnetised tape can be pulled past a similar coil – and a varying electrical signal set up in it (replay).

Poulsen, though, had no electronic amplifiers and the replay signal was too small to drive anything but very sensitive headphones. At the time, the acoustic gramophone was a far better recording and replay device.

Magnetic recording developed rapidly with the growth of electronic amplifiers, especially during the second World War. And during the fifties, there were many domestic **reel to reel** tape recorders around. But reel recorders were fiddly: the tape had to be laced round various guides and rollers before use.

So the German firm, Grundig, developed a **cassette** system – the tape was permanently laced-up round its reels and fitted into a box which simply slotted into the recorder. It didn't catch on – but a similar system developed in the mid-sixties by Philips and called the *Compact Cassette* did catch on, and in a very big way (partly because Philips allowed any tape and deck manufacturer to copy its invention, so long as they stuck rigidly to the Philips standard).

The new device, with its small size and ease of opera-

HOW TAPE RECORDERS WORK

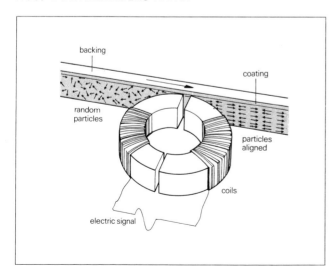

Tape consists of a plastic backing material with a coating of special metal oxide (sometimes pure metal) particles stuck on to it. The small particles can each be considered as a tiny bar magnet – with a north pole at one end, and a south pole at the other – capable of being pulled into a definite magnetisation pattern in a magnetic field.

To create a magnetic pattern, the tape is pulled across a *recording head*. In its simplest form, this is just an electromagnet – coils of wire wound on an iron core. The core forms a near-circle, except for an extremely small gap at the front; when a varying electrical signal is passed through the coil, a varying magnetic field is set up across this gap. This field is strong enough to pull the particles on the tape into a varying pattern which models the pattern of the electrical signal; the signal has been captured.

At a later date, the now-recorded tape can be pulled (at the same speed) past a replay head – essentially the same as a recording head, and often physically the same, too. The changing magnetism in the tape sets up a similarly-patterned changing electrical signal in the coil – which can be amplified and retrieved.

95

tion, was a natural choice for dictation machines, portable, battery-operated low-fi equipment, and in-car entertainment (where it has since more or less killed off the opposition – the 8-track cartridge player). But the small size and slow running speed seemed to rule out true hi-fi. That was left to reel-to-reel tape decks. One of the greatest problems was the intolerable level of hiss.

Then an electronics engineer, Ray Dolby, invented a clever system for significantly cutting down hiss on tape recording – the *Dolby System of noise reduction* (**Dolby** for short). The professional Dolby A System was followed by a domestic B version, suitable for (now hi-fi) cassette decks. (Virtually all cassette decks now have some sort of noise reduction system – usually Dolby.) Once Dolby had demonstrated that it was possible to get a quart into a pint pot, there was no holding back the inventors. The latest cassette decks, with their microscopically-fine mechanical engineering, and ultra-sophisticated materials in the heads and tapes, contain about a gallon of technology (and still they're pouring it in).

Domestic reel-to-reel decks hardly get a look-in these days, but the principles of recording apply equally to them. Information on reel-to-reel decks, and when you should contemplate buying one, are given in chapter 9.

Tape decks differ from record players and tuners in that they will record their own programme material, either live via microphones, or from gramophone records or a radio tuner (or another tape deck, of course). But, more than you probably realise, almost everything you would want to record is covered by copyright, or performing rights – which means it is illegal to record it, even for your own private use, unless you have permission. As this book goes to press, it is more or less impossible to get the necessary permission; a government green paper on the subject is expected.

Deck and tape matching

Getting good quality from a cassette deck is not so very difficult – but getting the very best quality can be rather a headache.

The main problem with any cassette deck is that the sound quality depends to a large extent not only on the *type* of tape used with it, but also on the *brand*: variations within the types, especially ordinary ferric tapes, can be very large. (The different tape types are explained starting on page 98.)

There would be little difficulty if intrinsically good tapes and intrinsically good decks would always give good sound quality when put together. But there are two technical reasons why this is not always so – to do with *bias* and *equalisation*.

Recording bias. For magnetic tape to be magnetised properly during recording, a very high frequency signal (the bias current) has to be passed through the recording head at the same time as the signal being recorded. Several things happen as the amount of bias current is increased from zero: the amount of harmonic distortion, the level of the low frequencies, and the level of the high frequencies all start to rise and then fall – and all at different rates. At very high bias currents, the tape becomes saturated and bad distortion also sets in.

It is important to get large amounts of signal on to the tape (since this determines the signal to noise ratio), while keeping acceptably low levels of distortion, together with balanced amounts of low and high frequencies (to get a flat frequency response). The amount of bias required to do this varies with the design of the cassette deck, *and* with the formulation (and therefore brand as much as type) of tape being used. So the amount of bias current put out by a particular deck will give best results only with the brands of tape that work best at that bias level. Indeed, even for a given tape and deck, there is no single bias current at which distortion, output level and frequency response are all optimised – so every combination is to some extent a compromise.

Equalisation. To get an extended high frequency response, large amounts of high frequency boost have to be applied. The amount of equalisation that is needed on *replay* to compensate for this has been standardised (if it had not been, pre-recorded tapes would play back with different frequency responses on different decks), with one level for ferric tapes and another for all other types of tape. But the amount of equalisation needed when *recording* varies not only with the design of a particular tape deck, but also with the level of bias used with any particular tape.

What you need to match. This all means that, for best results, you need to know what level of bias your tape deck provides, and at what level of bias each brand of tape works best. You need to use the brand of tape whose bias requirement most nearly matches the bias of your deck. If you get the bias match wrong, you may get poor performance from a tape which has sound qualities superior to the one that gives you optimum performance. As exact matching is not necessary, tapes and decks could be divided up into, say, five or six groups – perhaps given star ratings like petrol, so that you would know what range of tapes went with your deck. Unfortunately, decks and tapes are *not* marked in any way that will help you match them, so you have to rely on other methods.

How to match. The simplest method ought to be to rely on the tape deck manufacturer's recommendation. But these often leave a lot to be desired: they may have lists of conflicting tapes (ie brands that could not possibly work equally well), or include unavailable brands, or, sometimes, simply have the wrong choice. The only thing you really want to know is *what brand of tape was the deck set up, or aligned, using?* – and few manufacturers will tell you that, for a variety of reasons. One manufacturer used to produce a list of various brands of tape, indicating with a rating how suitable they were for his tape decks – but not any more.

Luckily, there are other guides. *Which?* indicates in its test reports what brand of tape it used in comparisons of cassette decks; most reviewers in Hi-fi magazines will at least try a range of tapes on each cassette deck they test. In *Hi-fi for pleasure* there is a *Tape check* feature each month, in which one or two tape decks are tested with a wide range of tapes, and the results ranked to show the best.

But do not take any of these guides as infallible. Samples of tape vary, and the alignment of decks varies, too, so the best tape for your sample of machine may be slightly different from that chosen by *Which?* or by reviewers.

This means that, in order to be sure of getting the best match, you have to do some **listening tests** for yourself. Record a section of a good record – one with wide dynamic range, plenty of high frequencies, and clean surfaces. Then replay the tape at the same time as you start to play the record again. By switching back and forth between the tape and the original record, using the controls on your amplifier, you can get a good idea of the quality of your tape recording. It is important to have the sound levels of the tape and record the same – so if your tape deck does not have playback level controls, then use an amplifier with variable input sensitivity controls on either the tape sockets or the disc input sockets.

Start by using a tape recommended by reviewers or *Which?* for your machine. If the dynamic range seems limited, and the high frequencies over-emphasised, then a tape with a lower bias requirement is needed. If the treble seems poor, try a tape that needs a higher bias.

But tapes are no better labelled than decks, so how do you find what brands have higher or lower bias requirements than the brand you are trying? Again, *Which?* reports on tapes list them in order of bias requirements, and some hi-fi reviews will give you enough information on bias requirements to help you pick.

Correct **mid-frequency sensitivity** is almost as important as optimum bias matching (the mid-frequency sensitivity is the level at which mid-frequencies play back relative to the level they were recorded at). Luckily, there is a fairly large degree of tolerance in matching – but if you notice too much or too little mid-frequencies in your listening tests, try a tape with a different sensitivity. Again, *Which?* reports and some reviews give information on the relative sensitivity of different tape brands.

Getting your deck fixed instead. A way of avoiding this game of hide-and-seek is to pick a brand of tape that is recommended as being intrinsically good, and then getting your tape deck re-aligned to suit it. This service is becoming more common among hi-fi shops and some manufacturers' service departments. It costs perhaps

CHAPTER 8

£20 or £30 – but you might be able to save this by having your deck aligned for a cheap, good-performance tape with different bias requirements from those your deck starts out with.

A couple of words of warning, though. A sow's ear of a cassette deck cannot be made into a silk purse by re-biasing for use with a good tape (though a silk-purse deck can become a sow's ear with the wrong choice of tape). And decks cannot be re-aligned for tapes with widely different bias requirements and sensitivities from the type it was originally set up for.

Not all shops and service departments will be equally good at re-alignment. When searching for a good firm, favour ones that have experience of this work, that are prepared to explain exactly what they are going to do, and that will try again if you spot problems (due to matching, rather than intrinsic tape or deck quality) in the disc against recording comparison.

Matching deck and tape is very important. Using the wrong brand of tape can literally turn the best-sounding deck on the market into one of the worst-sounding ones. Although tape and deck manufacturers give little help with matching, there are other sources of this advice.

Perhaps the best advice, though, is to get your deck aligned for a specific brand of tape. You could probably fund the cost of this by buying a cheaper deck than you intended to (you'll probably still end up with better performance), or by getting the deck aligned for a cheap but good tape, and saving on running costs.

What tape to buy

Cassette tape is less than 4mm wide, and almost always travels past the recording and replay heads at a speed of 4.75cm per sec (1⅞in per sec). The cassette itself is basically a slim box, about 4in by 2½in, in which the tape is wound on two reel hubs. The tape stays in this box all the time, even when it is being played or recorded on. The diagrams opposite show how it all works.

Most cassette tapes come in three lengths, which give different **playing times**. You can play or record with the tape running in one direction – then turn the tape over and play or record again with the tape spooling the other

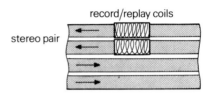

The stereo track layout on a cassette tape, with the left and right record/replay coils in the head. The tape travels from left to right, and the head records (or replays) the top pair of tracks; in effect, writing (or reading) the information from right to left. When the tape has all been pulled on to the right-hand spool, the cassette can be turned upside down. Everything reverses.

A mono recorder simply has one coil in the record/replay head, covering the full width of both stereo tracks. So both left and right channels of a stereo recording are read at once, and the stereo and mono systems are *compatible*.

way. The playing time quoted on cassettes is the total number of minutes of the two sides. Popular lengths are C-60 (at least 30 minutes playing time a side), C-90 (45min a side) and C-120 (an hour a side). The tape used in C-120 cassettes is rather thin, and some deck manufacturers do not recommend its use. Performance of one length may well be different from the same brand in other lengths. C-90 is by far the most popular length.

Types of tape. The first cassettes contained tape coated with an iron oxide. Most of the **ferric** tapes are now called low noise, or something similar – though this does not necessarily mean that they are particularly good performers. Ordinary ferric tapes are usually the cheapest type.

Then came the introduction of a **chromium dioxide coating (CrO_2)**. This was supposed to give better high-frequency performance, but many reviewers found chrome tapes performed rather poorly in other respects, particularly distortion. Often cassette decks are not properly aligned to take advantage of their benefits, either. The very latest chrome tape coatings (*formulations* in the jargon), have been getting better reviews.

Chrome tapes use a different amount of **equalisation** than ferric tapes. Most, if not all, decks have a switch to alter the equalisation; often, it is done automatically.

After the relative failure of chrome, the compromise **ferrichrome** tape was developed. This has two coatings: a lower one of ferric oxide, and a thin upper one of chrome dioxide. Performance is not outstanding – particularly mid-frequency distortion – and the tapes are relatively expensive. And again, one of the problems is that few tape decks are properly aligned to make the best of ferrichrome tapes. Modern ferrichromes use the same equalisation as chrome tapes.

HOW CASSETTES WORK

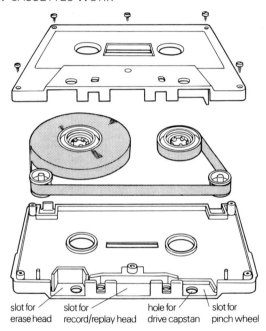

slot for erase head / slot for record/replay head / hole for drive capstan / slot for pinch wheel

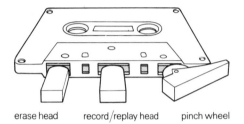

erase head   record/replay head   pinch wheel

The tape itself is wound on two hubs with holes in their centres for the drive spindles of a cassette deck. The pins facing into the holes mate up with slots in the spindles so that the spindles can drive the hubs without slipping. The tape passes round various guides and rollers moulded or fitted into the cassette housing; these differ from brand to brand.

There are three main slots in the front of the housing: from left to right, these take the erase head, the record/replay head, and the pinch wheel and drive capstan.

What happens when the cassette is inserted into a deck, and the replay button is pressed? Behind the tape, a capstan pokes through a hole in the housing; connected to the capstan is a large flywheel which helps it turn at a constant rate. This capstan provides the drive to pull the tape past the heads.

As the replay button is pressed, the erase and record/replay heads are pushed forward, through the slots in the cassette housing, and against the tape. A free-running, rubbery pinch wheel is also pressed up to the tape, opposite the capstan, so that the tape is sandwiched between the two. A pressure pad, built into the cassette housing, keeps the tape in snug contact with the record/replay head. As the capstan turns it turns the pinch wheel, and so pulls the tape from left to right across the heads. Drive linkages from the capstan motor (or separate motors in some cassette decks) wind the tape on to the right hand hub or spool as it is pulled through.

During fast winding or rewinding, the heads and pinch wheel stay out of contact with the tape, and the tape is wound on to the right hand or left hand hub by the cassette deck spindles.

The cassette has a couple of tabs on the back edge. If these are broken off, you cannot record on the tape, but you can still play back – so precious recordings can be protected against accidental erasure. When a tape with the tab broken off is inserted into a deck, a record locking pin drops into the tab hole – this prevents the record button from being depressed. If you decide that a precious recording isn't precious any more, a piece of sticky tape over the hole is enough to stop the locking pin from dropping into it.

Many chrome cassettes and decks have an extra hole and locking pin. When this pin detects the hole in a chrome cassette, it automatically switches the bias and equalisation from ferric to chrome type.

The so-called **pseudo-chrome** tapes are specially treated ferric tapes, designed to work with chrome equalisation. These tapes generally offer good performance, though you need a high-quality, well aligned tape deck to make the best of them. They can cost about a third as much again as the best good-value ferric tapes.

The latest arrival is **metal tape** – coated with a pure metal (iron) rather than a metal oxide. The problems with making pure iron tapes have been enormous – not least that the material tends to rust (ie convert into an iron oxide) until it is coated with binder and stuck on the backing.

The latest metal tapes and decks suffer from their own problems. One of the most serious is an exaggerated bias mis-match. Metal tapes require much more bias than even the highest-biased ordinary tape, so normal cassette decks cannot cope and new metal-compatible decks, with special designs of recording head and record and bias circuitry, are necessary.

An ordinary deck is likely to give *less* good results with metal tape than with an ordinary much cheaper tape. The gross under-biasing would give a harsh, distorted sound, and perhaps even worse dynamic range. Many ordinary decks may not be able to erase metal tapes properly, either, so on re-recording, the original programme would still be heard, faintly, in the background.

Of course, some so-called *metal-compatible* decks are less good than others – exhibiting the same problems as using metal tape on ordinary decks, though to a lesser degree.

There is probably nothing quite so confusing in the hi-fi world as the range of types and brands of cassette tape available. The solution to the brands problem is fairly simple: the brand you pick has to match your cassette deck (and preferably should be a good-value brand, too).

What type you should use is a little more difficult to decide. Probably the best solution would be to use a cheapish, good-quality *ferric* for everyday recording, and a very good *pseudo-chrome* for your more important or demanding recordings – good live broadcasts,

---

Cassettes versus discs – or do you need both?

The main advantage of a cassette tape system as a programme source is, of course, that you can record your own programmes – an advantage that records do not have. You can also use cassettes in car audio systems.

The sound quality of the best cassette decks and tapes is very high, especially if the recording source is of high quality – for example, live broadcasts on Radio 3. Purists, though, say that cassette quality does not and cannot equal that of the very best discs. Less critical listeners find it difficult to distinguish between a record and a good cassette recording of it.

A library of tapes that you have recorded yourself will be very much cheaper than an equivalent library of discs. (Though if you restrict yourself to legal copying, you will find opportunities for building a library rather restricted, as recording from disc and radio seem to be, strictly, taboo.)

Cassette decks and tapes are easier to operate (once set up and aligned properly) than a record player and discs. In particular, cassette tapes tolerate much less careful handling without damage than do records.

But a library of pre-recorded tapes will be rather more expensive than one of discs, and sound quality, though getting better, is rarely equal to that of discs. Pre-recorded tapes are not as widely available as discs, and not all works (or versions) available on disc are also available on cassette.

So, having a cassette deck and no record player makes sense only if either:

● you value their simplicity enough to pay the premium for pre-recorded cassettes; you aren't the most critical of listeners; and you want to buy only the more popular works

● or the cheapness of blank tapes is paramount, and you are prepared either to record illegally or restrict your library to material you may legally record (hardly anything at present).

say. At its best, *metal* tape is extremely impressive to a critical listener, giving a cleaner, clearer sound especially at high frequencies. But the tape needs a well-aligned metal-compatible deck to give its best – and in any case, the programmes you record (or your ears) may not be of sufficient quality to warrant its high price. But if you do decide to include metal, you should have the best results money can buy.

FEATURES ○|○|○|○|○|

Tape transport controls

These are the controls that set the tape moving – *play, fast forward, rewind, stop, pause, record, eject*. Many decks use large levers of the traditional piano keys style for these controls.

Some decks have **electronically controlled** tape transport. This cannot get rid of all the mechanical parts, and probably the greatest asset of electronic, logic-controlled and micro-switched transport functions is that they allow automatic control to be more easily built in. For example, some decks have a *tape search* facility: the deck will automatically wind the tape on, stopping and going into play at the beginning of the next recording. The system usually works by looking for blank passages between recorded items, and so may confuse quiet passages in orchestral music, or pauses in speech, for blank passages. One or two decks have *remote control* for the transport functions.

Whether mechanical or electronic, most machines *disengage the drive* to the tape when the tape is ended, saving strain on the tape and motors. Not all machines do this on fast wind or rewind, though.

If you are hunting for a particular passage on a tape, two features may help you (aside from a tape search facility). One is the ability to **change running modes without pressing stop** – that is, change among play, rewind and wind, directly. Many mechanical decks allow this (between some of the functions, if not all) and it is something that electronic-control decks find easy. The second feature is **cue and review**. The cue bit works during replay: you press the wind or rewind buttons; the play button remains depressed and you can hear, faintly, a speeded-up version of what is recorded on the tape. When you release the wind or rewind button, the machine carries on playing directly. The review bit works on record: after recording a passage, you can press the rewind button directly; the record button releases, but not the replay one; when you reach the beginning of the recording, you release the rewind button and the deck goes straight into replay. Both functions sound complicated, but can make life easier.

Decks with **timer stand by** can be set to start playing back, or (more useful) to start recording at a preset time. These are very crude compared with the timer functions on video recorders: for example, you have to buy your own timer (quite costly if accurate). And the short running time of cassettes – only 45mins uninterrupted with C-90s – makes unattended recording a little unattractive.

One way of extending the longest playing time is to get a deck with **automatic reverse**: when the tape has reached its end, the heads automatically move to the other two tracks, and the tape starts running in the opposite direction; and so on and so on. A deck with auto reverse only on playback, though, is not as useful as one with auto reverse on record.

Dual-speed decks are slowly coming in. The second speed may be twice the normal, or half it. Double speed gives better quality – but today's high-performing decks and tapes hardly need it (and playing time becomes extremely short). The half-speed option usefully extends playing time and makes the best use of high performance – though it will be some time before half-speed decks sounds as good as normal-speed ones.

Most recording tape has leader tape at the beginning and end – a short bit of plain plastic tape designed to protect the beginning of the magnetic tape from damage. Obviously, the leader tape has to be cleared past the heads before a recording can be started; the problem with most cassette decks is that you cannot see when this has happened. A neat device fitted to some decks is a **tape advance** button to clear the leader automatically.

Tape counters indicate the position that a tape has reached. Most are mechanical – rather like the mileometer on a car; a few have an electronic display for the numbers. Counters are a boon: without one it would be at best tedious trying to find a passage buried somewhere in the middle of 45 minutes of tape. But there are a couple of problems.

First, the counters are not linear: an increase of say 50 in the reading means different time periods, depending on where on the tape you start and finish. This can be true even if the counter display is electronic. If you need to know exactly how many numbers on the counter represent, say, each half minute of tape, at different points on the tape, you will have to draw up a calibration chart. Even if you do not go to these lengths, you will probably find it helpful to know what counter reading represents the end of the tape, so that (always supposing that you set the counter to 000 at the beginning of the tape) you will be prepared for the end while recording. Occasionally a deck comes along with rather better counter features – for example, with a **tape remaining meter**, scaled in minutes.

The second problem is that few counters are particularly accurate, and rewinding to a counter position previously noted may put you a few seconds off where you want to be. On the other hand, accurate counters are really needed only for the sort of recording that involves cutting and joining bits of tape, re-recording and so on – activities usually confined to reel-to-reel decks. So the counters supplied with cassette decks are usually adequate.

A **counter memory** is quite useful, however. You set the counter to 000 at any point on the tape (at the beginning of a particular recording, say), and the tape will automatically stop there during rewinding.

Tape type controls

The different types of tape require different amounts of equalisation. They also require different levels of bias, and, at least within ferric types, each *brand* ideally requires its own bias level for best performance. Most machines therefore have **switches** to alter bias and equalisation.

The most flexible arrangement would be separate switches for bias and equalisation. Replay equalisation is standardised, and there are only two levels – one for ferric tape, and the other for chrome, ferrichrome and metal. So a two-position switch is all that is necessary.

Bias switches can be rather more complicated. To cover the whole range of tapes, a six-position switch could be needed; with positions for metal, pseudo-chrome, chrome, ferrichrome, and two positions for different types of ferric tape.

Alternatively, a single ferric position could be provided, along with a variable bias control. For a particular brand of ferric tape, the best position could be found by making recordings of a good disc and playing them back, altering the bias control with each test recording until the distortion levels and frequency responses of disc and recording appear to be the same (see page 97).

With some decks, particularly three-head ones (see opposite), there is **automatic bias adjustment**. A low and a high frequency tone are recorded from built-in generators, and on replay (which can happen at the same time on a three-head machine), the output is displayed on the recording level meters – one tone on the left meter the other on the right, for example. The bias control is adjusted until the two meters read the same. However, all this tells you is that the output at these two frequencies is equal; you have to assume that the response in between (and above, for the high-frequency tone might be as low as 10kHz) is flat. You might find manual tweaking gives a result that you consider better.

The same applies to **fully automatic bias and sensitivity setting**. A tape is placed in the machine, and shuttles backwards and forwards until the machine's microprocessor is satisfied that it has optimally set bias and sensitivity levels. How well it does this, of course, depends on how sophisticated the automatic circuitry is.

All this sounds very complicated, and many decks have much simpler tape type switches. And unless you want to experiment with a wide range of tape types and brands, simple tape switches *with the electronics properly aligned for the tapes you are using*, will be quite sufficient.

Number of heads

All cassette decks have at least two heads. One simply erases tapes just before they are re-recorded. The other doubles up as a recording head and a replay head. This theoretically limits sound quality, since the technical requirements for a good recording head are different from those for a good replay head. It is also less convenient from an operating point of view. For example, it is not possible to listen to what has been recorded while the recording is being made. Of course, you can listen to the input to the tape – what is *going to be* recorded – but you have no idea of how well the tape has actually been recorded until later.

So a **three-head machine** (one with separate erase, record and replay heads) sounds like a good idea. And it probably is, operationally. For example, as long as the replay head is mounted after the record head (usually on its right), true *monitoring* can be carried out. This means that the signal that is being recorded by the record head is picked up a split second later by the replay head. So faults – perhaps distortion caused by recording at too high a level – can be quickly remedied, and only a small portion of the recording will be poor. You can also check the recorded sound and the original almost instantaneously, while the recording is taking place – very useful when setting up variable bias controls, or checking different tape brands.

But how well separate heads *perform* depends on their design and the ingenuity shown in shoe-horning them both into the gap originally meant to take only one head.

Three-head decks can be a useful feature, particularly if you spend a lot of time making critical recordings or trying different tape brands, but they may not provide better performance than a two-head machine.

Recording level indicators

It is essential to know how much signal is being recorded on to a tape. Too little signal means that the signal to noise ratio will be poor (ie, on play back there will be more hiss than is necessary). Turning the recording level controls up to increase the level of the signal being recorded will give a better signal to noise ratio, but (beyond a certain point) only at the expense of more and more distortion. Recording level indicators show how much signal is being recorded, and so will help you select the level of signal that gives the best compromise (for you) between distortion and signal to noise ratio.

Most decks have a pair of **recording level meters** – often called VU (for volume unit) meters. Most are scaled in dB, with 0dB usually being about two-thirds of the way along the scale. The best compromise between distortion and signal to noise ratio is supposed to be when the level controls are altered so that the meters' needles do not rise above the 0dB mark.

But the story is much more complicated than that. For a start, most meters over-read on steady tones (that is, distortion does not become audible until the needle is well over 0dB). So keeping the needle at 0dB or lower results in poorer signal to noise than is necessary. On the other hand, most meters do not react quickly enough to sharp peaks or transients – so it under-reads on these, and attempting to record peaks reading 0dB could result in large amounts of distortion.

Then, the amount of recording signal that a tape can handle before giving audible distortion on replay, and the signal to noise ratio on replay that a particular recording signal will produce, depends on the brand and type of tape being used. Hence the best position for the meter needle depends on the tape, too. Finally, the best position depends also on whether it is distortion or noise that bothers you more.

This all means that the only way to ensure you are getting the best distortion and noise performance from your combination of deck, tape and ears, is to make some trial recordings of the type of music you are likely to record, with the recording level controls in different positions, until you find the most acceptable compromise. This doesn't mean that the meters are redundant: when you find the best compromise for you, note the readings on the meters (during recording, not playback) both during steady bits of music, when the needles are hardly moving, and during peaks, when the needles will be flapping about violently. Then, whatever source you are recording from – record deck, tuner or whatever – if you set the level controls so that the loudest peaks and steady passages do not exceed these readings, then your recordings should be as good as you can make them.

If a meter gave the same reading on transients that it did on steady passages, at least one complication would be removed. Some decks do have **peak reading meters** (or meters that can be switched to read peaks), and most now have **peak reading lights**, which light up when a particular signal level has been recorded.

Assuming the peak meter is working well, **peak hold** is useful: the needle stays at the level of the highest peak encountered for a few seconds, before falling back. This gives you time to read the level easily, which a rapidly-flickering needle would not.

It seems to be a rule that electronics displaces mechanics wherever possible, so some decks have meters made up from a row of lights, or a fluorescent bar that changes length. One deck has five rows, each covering a different part of the frequency range. In theory, this is useful, because the level of signal a tape can accept is different at different frequencies, and the frequency content of music is different between types, too. But in practice, you may find such a display just more difficult to interpret than a single row of lights or a meter needle.

Good metering is essential for good recording. But whatever system is used, you have to spend some time experimenting to find the recording level that gives the best results, and learning how to interpret the meter readings accurately.

Noise reduction

One of the most important problems with any recording system – and cassette tape in particular – is getting a good dynamic range, so that the quiet sounds are not lost in hiss, and the loud sounds do not overload and cause distortion. Without doctoring, a cassette recording system has a dynamic range of roughly 55dB – nowhere near enough to cope with some music.

The solution is, essentially, very simple. There is really no reason why a recording should be directly a representation of the original sound. For example, a 3dB change in level on the tape need not represent a 3dB change in the real music. If the music is *electronically compressed* before it is recorded, the recording will have a smaller dynamic range than the original. For example, if the compression ratio is 2:1, then a 70dB dynamic range in the music will occupy only 35dB on the tape, and so will sit well above the noise level and below the distortion level. On reproduction, the music is expanded by the same ratio to restore the original dynamic range.

As usual, what sounds fine in theory does not always work as well in practice. One problem is that compression and expansion (*compansion*, in the jargon) can highlight frequency errors. An error in frequency responses of $1\frac{1}{2}$dB between the recording and the original, which may well have passed unnoticed, could become a 3dB error (much more noticeable) if 2:1 compression is used. Another problem with some types of system is *noise pumping*: the level of noise can be heard to increase and decrease, according to the loudness of the signal. Some listeners may find this more distressing than steady noise, even if this is louder.

Certainly the most famous and important noise reduction system (as well as the first domestic one) is **Dolby B**. This applies compansion only to quiet, high-frequency sounds, since it is with these that noise is most apparent. The level of noise reduction (another way of saying the amount of increase in dynamic range) is not very great (about 10dB), but the system is reckoned to be quite free from side effects. It does need careful setting up, though – for one thing, to suit the sensitivity of the tape being used.

The JVC **ANRS** (Automatic Noise Reduction System) is very similar to Dolby B. But the **Super ANRS** (SANRS) works in a slightly different way, by compressing and expanding high frequency transients; some experts find this works less well on material such as piano.

There are two main compansion systems which work over the whole frequency range, and at all signal levels – **dbx** and **Adres**. These can offer an enormous degree of noise reduction (30dB or more) and do not need the careful setting up of systems like Dolby B. But some listeners may find the sound quality rather poor on some types of material. There are few cassette decks available with dbx or Adres built-in, but dbx is available as a (quite expensive) add-on unit.

High-Com works in a way that is somewhere between

Dolby and dbx. Some reviewers find it doesn't perform as well as Dolby.

New systems. The demand for a greater amount of noise reduction with good quality sound has prompted Dolby to bring out two new systems, which are just starting to become available on cassette decks. **Dolby HX** varies the bias signal according to the frequency of the signal being recorded, reducing it for the high frequencies, but leaving it at a higher level for the mid and low frequencies. This allows greater amounts of high frequency sound to be recorded, but still keeps mid and low frequency distortion acceptably low. Tapes recorded with only Dolby B noise reduction can still be played back on decks with HX, since the HX circuitry works only on the record side, not on record/replay like most other systems. HX is used as well as, not instead of, Dolby B.

Dolby C is, very roughly, two Dolby B units one after the other. The net effect is a noise reduction of about 20dB – twice as good as Dolby B and, when used with the best tapes, about as good as necessary even for the strictest listener, the highest quality recording, and the most difficult material. Dolby C is designed to be compatible with Dolby B, and most experts say there is no more audible effect on sound quality than with Dolby B.

Some pundits say that, with the advent of digital sound – see page 22 – noise reduction systems will become redundant. But they should still be useful – for example, to get good sound quality from analogue cassettes running at speeds lower than 1 7/8ips, or from the smaller micro cassette sometimes used in dictation systems. It might also be possible to link some type of noise reduction to a restricted digital system – for example, to lower the tape speed used.

Dolby B noise reduction (or something very similar) is essential in present day cassette decks. From the point of view of hiss reduction, the new super systems, such as HX or C, are about as useful as changing to metal tape – a wise choice if you want the ultimate sound, but not absolutely essential. On the other hand, decks with the new systems, when available, are not likely to be much more expensive. One of the new systems (and, possibly, metal tape as well) will be essential for good quality on slow-running decks or micro-cassettes.

Level controls

It is essential to be able to vary the amount of signal that is fed to the recording head, so all decks have **recording level controls**. The only difference between them is the type. Easiest to operate are *ganged knobs* (one inside the other, on the same spindle), or *sliders* set near enough together so that they can be operated with two fingers of the same hand. Perhaps a little less easy would be a single stereo knob or slider with a separate balance control. A few decks have separate, unganged knobs, which require two hands to set the level. This is awkward: it is useful to have a hand free for operating the transport controls.

Playback level controls, on the other hand, are not essential. Their main effect is to alter the level fed to the amplifier – but most decks and amplifiers can cope with a wide variation in input levels (at least, with phono sockets: DIN sockets are something different, but playback level controls may not alter the level from DIN sockets, anyway).

They are useful for setting up decks and tapes, though (see page 97), and some alter the level fed to the headphone socket, which can make listening on phones more comfortable – again, though, something most amplifiers have.

Look for recording level controls that are easy to operate one handed. Playback controls can be useful, but are not essential.

MPX filter switch

MPX filters cut down the level of pilot tone and subcarrier frequencies in a stereo radio broadcast. These can cause interference with the taping, and with the Dolby circuits. So MPX filtering has to be included in all Dolby systems. Often, though, the tuner itself adequately filters out these frequencies. In that case, if your deck has an MPX filter switch, you might get a better frequency response without the filter.

CHAPTER 8

Sockets

The usual arrangement is *phono sockets* for input and output (and generally a combined *DIN* socket as well), mounted on the rear of the deck. Sockets for *headphones* and for *microphones* (a single stereo jack socket for the phones, and a pair of mono jacks for the microphones) are mounted on the front.

One or two decks mount a second (usually DIN) input/output socket on the *front*. This would be useful if you spent much time plugging and unplugging various pieces of equipment – though, since the second socket is usually DIN, the advantage isn't very great, and you would probably be better off buying separate connecting units that you could tailor to your particular requirements.

Portable recorders

Most portable cassette recorders are, of course, low-fi – often, these days, part of a radio-recorder. There are few hi-fi portable cassette recorders available.

Some hi-fi recorders *can* give you sound quality as good as mains equipment, together with portable operation from batteries. But they are generally expensive. Consider carefully whether you need the portability – and if you do, whether you need the hi-fi quality as well; it might be best to buy a good mains machine, and a cheap low-fi portable.

DECK TESTS ☑ ☐ ☐ ☐

Cassette deck and tape tests are still evolving – not surprisingly, considering the rapid changes in the technology. The main tests (on both decks and tapes) are to do with *frequency response* and *signal to noise ratio*. There are also mechanical tests, such as *wow and flutter*, and tests to evaluate how the *meters* behave. *Distortion* tests usually form part of the signal to noise ratio measurements.

As cassette decks and tapes are used together, it makes sense to test them together: testing each part separately, and then trying to predict the sound quality of the combination is fraught with difficulties.

But testing each combination of deck and tape is cumbersome and costly, and, if only in order to draw up a short-list of tapes likely to suit a particular deck, tests on tapes themselves (see page 109) are important.

Frequency response

One of the main problems of cassette decks (aside from hiss) is that they are unable to handle large amounts of high-frequency sound. Luckily, a lot of music does not contain high levels of high frequencies – but when it does, and the decks fail to handle it, the sound can be very unpleasant. Brass, for example, sounds dull and squashed (the effect is called *hf compression*, or *hf squash*). Manufacturers of decks and tapes have put most of their effort into improving the performance at the high frequency end – ensuring that tapes will handle larger amounts of high frequencies before compression sets in.

Frequency response measurements usually ignore hf compression by using a relatively low input level of minus 20dB or lower. (The input level can be referred to 0dB on the deck's meters; check on your deck to see how low a level minus 20dB is, when compared to the levels you normally record at – or to one of the flux reference levels – see Box.) At low input levels, and with a correctly-biased deck, a response flat to 14kHz or so would be a good result. Many decks show a slight fall-off in response at low frequencies; so long as this is confined to below 50Hz or so, it is probably not serious.

But listening tests for *Which?* reports have shown that a good response at low input levels is not all that is necessary for good sound; hf compression has quite a large effect. With some combinations of deck and tape, compression can be noticeable even at low input levels.

Dolby noise reduction circuits also affect frequency response. They work by assuming that the signal on replay will be at the same relative level as the signal on record. This would be so only if the tape sensitivity (ie the output from the replay head for a given recording signal) was the same from brand to brand. Large differences in sensitivity will cause Dolby *tracking errors* – and since the effect of the Dolby circuits varies, depending on the frequency and level of the signal, the frequency response will also show variations from flat which will differ at different frequencies and input levels. Fortunately, the system is fairly tolerant, and only large differences in sensitivity will cause noticeable errors; even so, large differences do occur between some tapes.

Many specifications show frequency response plots with Dolby *off* so that you cannot see the effect of any Dolby tracking errors.

Flux levels

For measurements and testing, it is very convenient to have a standard level of tape magnetisation, to which all other levels being recorded can be referred.

For cassette decks, there are two main levels. The *Dolby level* is equivalent to a flux (that is, an amount of magnetisation on the tape) of 200nWb/m (nano Webers per metre). On a deck replaying a tape recorded to Dolby level, the meter needles should point to the Dolby symbol, which is usually a few dB above the 0dB mark.

The *IEC* or *DIN level* is a little higher – 250nWb/m when measured by the DIN method; about 235nWb/m when measured by the same method used to establish the Dolby level. So in practice, the DIN level is about one and a half dB higher than the Dolby level; often the difference is unimportant.

Replay frequency response and azimuth. Most tests on cassette decks are done by recording on a tape in the machine being tested, and then playing back that tape on the same machine – record/replay tests. This makes sense because most people use cassette decks to play back recordings they have made on that machine, rather than to play back pre-recorded cassettes. Record/replay tests are often more stringent than replay-only, too.

But measuring the *replay-only* frequency response as well is useful for two reasons. First, it does give some idea of how pre-recorded cassettes will play back. Second, it tells you about *azimuth*, the angle that the gap in the recording head makes with the tape. For best high-frequency response, this should ideally be at exactly 90° to the length of the tape, but performance will not suffer if the angle is nearly correct, as it is exactly the same both on record and replay. So for tapes recorded and replayed on the same machine, azimuth is largely unimportant and, unless it is grossly wrong, will not be noticed. But a pre-recorded test tape is made on a machine whose azimuth is very carefully set up at exactly 90°. Replaying this on a deck with poor azimuth adjustment will result in a loss of high frequencies. Quite a few of the cassette decks that *Which?* test arrive with poor azimuth, and so give a poor replay-only frequency response. When adjusted (which is really a job for a service technician), most decks then give at least a reasonable response.

Do not take too much notice in reviews and specifications of replay-only frequency response. For a start, it is likely that the variations in response (and azimuth alignment) of pre-recorded tapes themselves will influence the result you get in practice. However, if you are interested in listening to pre-recorded tapes a lot, then it is sensible to get a service technician to adjust the azimuth. This will give you the best chance of getting good high frequency response.

Three-head machines are a special case, since the record and replay head gaps have to be exactly parallel to get a good frequency response even on record/replay. Some machines have the record and replay heads mounted together, so that alignment between them is fixed during manufacture. Some decks with separate heads have a user control to adjust alignment, together with some means of indicating when the azimuth is correct.

Signal to noise ratio

On cassette decks, this measurement is often known as **dynamic range** – and indeed, the maximum dynamic range is the figure that is most worth knowing. As usual, it is the ratio of the loudest sound that can be recorded (for a given amount of distortion – usually 5 per cent third harmonic distortion) to the background noise.

Most reviewers now use CCIR/ARM/2kHz. These conditions result in a fairly large number for signal to noise ratio, so it is unlikely that manufacturers' specifications will use any more stringent method. Watch out, though, for noise measured unweighted (which might make the result appear a couple of dB better). Cassette decks are significantly less noisy than they used to be: a good result would be about 64dB on ferric tape; possibly 2dB better on pseudo-chrome. There is still room for improvement, though: a recording of a very good radio broadcast will still sound a little hissier than the original, even on a very good deck.

Signal to noise measurements are often also made using a lower signal level, hence giving a poorer-looking result. The signal is often a tape recorded to the standard DIN or Dolby reference level. A good result for noise referred to Dolby level would be about 60dB with ferric tape.

As with other types of equipment, hum, as well as hiss, can be a problem.

Distortion

For those more worried by distortion than hiss, measurements of distortion might be interesting – but there are problems with these. The amount of distortion depends first on the *level* of the signal. DIN or Dolby reference levels are commonly used, and these are probably about as typical as any other level.

Second, the *frequency* used for the measurement is important: a deck may give good results at the standard frequency of 333Hz (usually the third harmonic of this is measured), but may give poor results at high frequencies (around 10kHz) as a result of hf compression – especially if the deck is overbiased for the tape used. (Overbiasing in itself produces a poor frequency response.)

Third, distortion on *transients* is important – again, perhaps because of hf compression.

Many decks will give less than 0.5 per cent third harmonic distortion (at Dolby level). Anything below 1 per cent may not be audible. But listening tests are really needed to check on audibility, and to check distortion on transients.

Meters

As page 103 suggested, although good metering is crucial, there are many factors that can influence what the 'correct' reading can be. So measurements of meter accuracy are of limited use. What is important, is to know how to interpret them for the sort of tapes and type of music involved.

Separation and cross-talk

Stereo separation, as usual, is a measure of how well the deck keeps apart the left and the right channel signals. Cross-talk in tape decks, although often used synonymously with separation, is a measure of how well the deck keeps apart signals on side one of the tape, from those on side two.

There should be no problems with cross-talk, if the heads are aligned properly, and there is rarely any problem with stereo separation, either.

Wow and flutter

Wow and flutter vary depending on the cassette and the position on the tape where the measurement is made. So it is usual to make measurements at the beginning, middle, and end on each of say three cassettes, and calculate a fair wow and flutter figure from all the readings.

Less than about 0.1 per cent wow and flutter DIN peak weighted, or roughly 0.05 per cent WRMS should not be noticeable, and anything less than about 0.2 per cent is a good result for a cassette deck. Although the DIN peak weighting correlates quite well with the subjective annoyance of wow and flutter when the amounts

are large, there is some evidence that it is less accurate when the amounts are small. So a listening test, using material having long, sustained piano, woodwind and brass notes in it, is important.

Speed accuracy

Absolute speed accuracy is even less important on tape decks than it is on record decks – except for pre-recorded tapes. So long as the machine plays back at the same speed it recorded, speed accuracy is not important at all.

In any case, few decks have a worse accuracy than about 1 per cent, which shouldn't worry even the most fastidious listener. The occasional deck may take a few seconds to run up to speed, which is annoying. Before making important recordings, let the deck run for half a minute or so, and use the pause button.

Tape guidance

Unless the recording tape is kept firmly in contact with the heads, the output will vary slightly in level. Large, momentary drops in output are called *drop-outs*. By replaying a standard test cassette, the consistency of the output can be checked, and the level, length and frequency of drops-outs noted. Obviously, the more consistent the output, the better the head-to-tape contact – but exactly how this correlates with what is heard is not known. Again, the overall sound quality in a listening test is more important.

TAPE TESTS ☑ ☐ ☐ ☐

Optimum bias

The results from bias measurements are only relative: although the bias voltage produced by the test deck's electronics can be measured, the result has no meaning for any other deck. So all that can be said is that a particular tape has an optimum bias requirement so many dB above or below another brand of tape. With experience, though, this is enough to be able to decide whether the tape is likely to be suitable for a particular deck or not, even if the exact results of using it cannot be predicted.

Dynamic range

The tape's own dynamic range can be measured in the same way as that of the combination of deck and tape. The result can tell you whether the dynamic range of the combination is limited by the deck – in which case changing to a better tape will not help – or by the tape. But the results for tapes generally do not allow for the improvement that Dolby noise reduction will bring – between 9dB and 10dB, depending on the deck.

The largest signal that can be recorded (without giving more than 5 per cent distortion on playback) also determines the greatest output from the tape. This is called the *maximum output level* – MOL for short. MOL is commonly referred to Dolby or DIN level; good tapes can produce an output at mid frequencies about 6dB higher than Dolby level. Although this does not tell you much about dynamic range (unless you know how much below Dolby level the level of background noise is), it is still a useful measurement. Most improvements in tape dynamic range and frequency response come from improvements in MOL at high frequencies – so monitoring these gives a good idea of the quality of the tape.

The recording level meters on a deck will have been adjusted for a tape with a specific MOL. So if you use a tape with a higher MOL, you will need to let the meters read higher than previously if you want to take full advantage of the tape's dynamic range.

MOL is a useful guide to various properties of a tape. But be wary about equating it with dynamic range: although a high MOL and a wide dynamic range generally go together, in a particular case, a tape with a relatively low MOL but very low noise may have a better dynamic range than another with higher MOL, but high inherent noise.

Tape sensitivity

This is the amount of input signal required to put a given amount of magnetisation on the tape. Differences in sensitivity can affect frequency response with Dolby on (see page 107) – so if you are experimenting with different tapes, try to use one with much the same mid-frequency sensitivity as the tape the deck is adjusted for (or get the deck re-adjusted). Again, absolute values of sensitivity mean nothing: it is only the level relative to other tapes that is important.

A sensitive tape is not inherently better than a less-sensitive one; the only difference is that you will need to turn down the recording level controls a little to get the same output.

Frequency response

The optimum bias setting is usually chosen to give as flat a response as possible – so a frequency response plot is unnecessary. Of course, if the tape is tested at other than its optimum bias (or if a different definition of optimum bias is used) then the response might not be flat. But even in this case, the result in practice will depend as much on the deck used as the tape.

Print-through

Recorded cassettes are often left for long periods without being played. Particularly if they are kept in a warm place, magnetisation can be transferred from one layer of the spooled tape to adjacent layers – *print-through*. The effect is to cause echoes on the tape and, although the adjacent layers are only weakly magnetised by the print-through, it can still be audible, particularly if a loud passage happens to lie adjacent to a quiet passage. You may hear the echo before the actual passage (*pre-echo*) or after it (*post-echo*). In orchestral music, it is likely to be the pre-echo that is heard – in the silence that often precedes a loud chord. And there is often a silent gap just after the passage of loud speech – where post echo might be heard.

Print-through can be measured by recording loud pulses – say 2sec on and 10sec off – rewinding and storing the tape for a few days, then replaying it and measuring the strength of the echoes that the print-through has created. This is a stringent test, and a good result would be a level of echo about 53dB below the level of the pulse.

Drop outs

Head to tape contact can be measured for tapes in the same way that it is measured for the combination – see *Tape guidance*. But again, correlating this with what is heard is not easy.

Mechanical tests

Some manufacturers make much in their advertising of particular mechanical features of their cassettes which are claimed to make for smoother running, or to reduce the risk of jamming, or to lower wow and flutter. Wow and flutter can be measured – but there is little difference between brands, and most wow and flutter is due to the cassette deck, and not the cassette. However, as decks improve, the performance of the cassette itself may become more important.

It is sensible to wind a cassette back and forth before making an important recording, if you are worried about the possibility of it sticking. These days, nearly all cassette housings are screwed together, so it should be possible, with care, to take them apart and unjam the tape, if it ever does stick.

Tape deck care

There doesn't seem to be as much gadgetry on the market for keeping tape decks clean as there is for discs. This is surprising because, in some ways, the effects of dirty or misaligned tape decks are more noticeable than similar problems on discs.

The main problem is that even the best tape will lose a little of its coating as it rubs across the tape deck heads. These fine particles of oxide will get just where they are not wanted – into the minute gap at the front of the heads. The result is muffled sound. Regular and frequent cleaning of the heads, therefore, is essential if you want to preserve the treble response.

Probably the best cleaner is a **solvent**, either one sold specifically for the job, or isopropyl alcohol from a chemist's shop. Apply a small amount with something like the end of one of those padded cleaning sticks sold for cleaning children's ears. At the same time, you can wipe away oxide deposits from any bits of the mechanism that are visible. Never use force, or a metal tool, or any other cleaning fluid.

An easier method is to use a **cleaning tape** – an ordinary-looking cassette filled with a tape designed to clean the heads. Some brands of recording tape have a length of cleaning tape at the beginning, so that the heads are cleaned automatically, every time you play the tape. Even with cleaning tapes, you may still want to resort to cleaning by hand occasionally.

Tapes themselves, unlike discs, are not cleaned, and do not require the sort of careful handling that discs do. But it is important to keep them away from strong magnetic fields, such as the one inside a loudspeaker, or you may find they lose the recording made on them. Keep them away from heat.

Chapter 9 Reel-to-reel tape decks

Ten years ago, if you wanted to add tape-recording to your hi-fi, you would have bought a reel-to-reel tape deck. Then, there were lots of budget-priced models to choose from. Now there are none at this end of the market; everyone buys cassette decks. For serious recording, however, reel-to-reel decks still have a place, even though the cheapest available costs about four times as much as a budget cassette deck and the most popular models cost between £800 and £1,000. This chapter tells you how to decide whether you should spend this much on tape-recording facilities, or whether a cassette deck would do everything you would want.

Cassette v. reel-to-reel – pros and cons

To help you decide what sort of tape recorder you want, consider first their vices and virtues in different areas.

Playing time. Cassette deck playing time is very limited. Leaving aside C-120 tapes, which are rarely used, the normal maximum is just 45 minutes uninterrupted. This could be doubled with an automatic reverse deck (one that reversed on both replay and record), or with the use of half-speed decks – but these features are found in very few machines. Reel-to-reel playing time is much less limited: two to three hours uninterrupted is quite possible.

Speeds. Almost all cassette decks have only one speed. A few have a second speed, though quality at a lower-than-normal speed is not yet very high, and playing time at a higher-than-normal speed is very short. Most reel-to-reel decks have at least two speeds, often three – and there are four or five speeds in all to choose from. Having a variety of speeds is useful: the faster the speed, the better the quality can be; the slower the speed, the longer the playing time, and the less the tape costs are. You can decide what trade-off of one for the other you want.

Track formats. Reel-to-reel decks offer the choice of *quarter-track* (reversible) or *half-track* (one-way only) operation – see page 114. Quarter-track offers greater tape economy; half-track can offer better sound quality and less need for critical alignment of the recording heads. With cassette decks, you have no choice: you're stuck with quarter-track.

Space. Cassette decks and tapes are small compared with sometimes quite massive reel-to-reel decks and their tapes.

Tape costs. Cassette tape is cheaper – around £1 an hour – than reel-to-reel, which can cost as much as £10 an hour, depending on the speed and track format used.

Convenience for normal use. The cassette medium was designed to be easy to use; the cassette tapes simply slot in and away you go. By contrast, reel tape is fiddly.

Convenience for creative use. Creative tape recording usually involves cutting out and re-ordering small lengths of recorded tape – editing. It also involves being able to use a deck easily and flexibly in live recording sessions. This is where reel-to-reel decks (especially 2-track ones) score. Many of them have facilities for editing built in, as well as other features to make live recording easy; and all have three heads for proper monitoring. Using a cassette deck for creative recording is almost impossible: it is difficult to cut and splice cassette tape accurately, and other editing facilities are very sparse. Few cassette decks even have the three heads needed for proper monitoring.

Pre-recorded tapes. Only very specialised material is ever available on pre-recorded reel-to-reel tape.

Sound quality. It is getting very difficult, even for expert listeners, to hear much difference in sound quality between the very best cassette decks and tapes and even studio-quality reel-to-reel. For most people, the differences would probably be irrelevant.

Noise reduction. A good noise reduction system, such as Dolby, is essential on cassette decks, and is included in them all. Noise reduction is rarely built into reel-to-reel decks. Yet, unless you use a very fast tape speed, some form of noise reduction is necessary on reel-to-reel decks to get hiss below the level of that on a good cassette deck. Add-on Dolby B units (the same type as is used in cassette decks) are rare, and would probably add £80 or more to the price of the deck.

Tape/deck matching. To get the best results from the cassette medium, it is essential to use a matching brand of tape with your deck. On reel to reel decks it is less important to get matching exact.

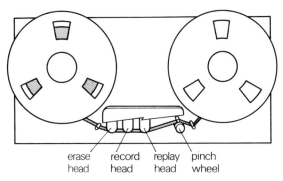

erase head   record head   replay head   pinch wheel

How tape is laced up on a reel-to-reel deck. The tape moves from left to right, emptying the supply spool, and filling the take-up spool. Unlike cassettes, the threading of the tape across heads and between rollers onto the supply spool has to be done by hand – with some decks, very fiddly. Note that the heads are on the *opposite* side of the tape from those on cassette decks; reel tape is wound with the magnetic side inwards.

To sum up, there are really only three reasons for getting a reel-to-reel deck rather than a cassette deck:
● **you need the extended playing time**
● **you need to be able to edit tapes, or make lots of high-quality studio-type live recordings**
● **you have very critical hearing, and find that cassette decks cannot match the quality of the best reel-to-reel deck.**

The price you have to pay for a reel-to-reel deck is high – probably four times or so as much as for a good cassette deck. So unless you are sure one of the three reasons applies to you, you'd be better sticking to cassette.

FEATURES ○|○|○|○|○|

Although many of the features on reel-to-reel decks are the same as on cassette decks, there are some differences – and the relative merits of different features varies depending on what you intend using a reel-to-reel deck for. In this section *domestic use* means recording from the radio or disc, or for the occasional live session. *Studio use* means complex live recording and editing work.

Tape transport controls

These are much the same as on a cassette deck, though there is more use of electric or electronic control. This isn't a gimmick: full reels of tape are very heavy, and if they are to be spooled neatly and to the correct tightness, and if you want to be able to change running modes without tape spilling out disastrously, it is important for the tape transport to be fairly sophisticated. With electronic logic control, it is likely that the deck will be able to offer timer facilities – useful with the long playing time – remote control functions and, perhaps, synchronisation with a slide projector. Some decks have variable fast forward and rewind speeds.

For studio work, it is important that the controls are easy to use without fumbling; for domestic use, this is less important.

CHAPTER 9

Speeds, track formats and spools

There are four main **speeds**: 76, 38, 19 and 9.5cm per sec (30, 15, 7½ and 3¾in per sec). By contrast, the standard cassette speed is 4.75cm per sec.

The highest speed is really used only by professionals but the other three are all available on semi-professional decks. A few decks have all three speeds; more likely, the deck you want will come in two forms – one with 9.5 and 19cm per sec speeds; the other with 19 and 38cm per sec. Choose the lower speed version if you want the maximum uninterrupted playing time (for most domestic uses) and the higher pair of speeds if you want to make the highest quality recordings (generally studio use).

There are two main **track formats** – quarter-track and half-track – see below for details. Go for quarter track for economy, half-track for the best performance. In fact, with the very best machines, there is little to choose in performance, but half-track would still be the best choice for studio work – much better for editing, and adjustment and alignment not quite so critical.

As with speeds, you will find that some decks are offered in two forms, with either quarter-track or half-track operation. There are a few decks on which you can change the head block, and you may find the odd deck which is switchable between quarter and half-track – though if these facilities are for replay only, they are of little use.

Note that quarter-track does not mean the deck is capable of recording or playing back discrete quadro tapes. For this, you would need a **four-track** machine (a term sometimes used wrongly for a quarter-track machine). This could also be of use in studio work for stereo recording: each track could carry the output of a different microphone, and the tracks could be amalga-

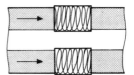
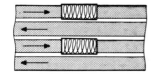

Half-track layout on reel-to-reel tape and right: quarter-track layout, with dimensions. On the half-track format, the tape passes through only one way – you cannot reverse it and record again (if you try replaying a half-track recording inverted, it will come out backwards). On the other hand, the recorded areas and the guard bands are much wider than they are for the quarter-track format – so the recording tends to be less delicate and more tolerant of incorrect head alignment; the signal to noise ratio should be better, too.

The quarter-track format works as for cassette tape (compare page 98). But note that the track configuration is different: the left and right channel of each stereo pair interleave with each other, rather than lying side by side. This should give better channel separation, but would not be stereo/mono compatible. (Playing back a stereo tape on a mono machine would give one channel, plus a backwards version of one channel on the other side.)

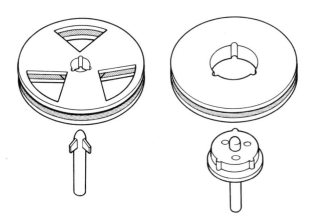

The normal hub on reel spools is the one used also on cine film – so it's called the *cine hub*. The three slots in the spool's centre fit over three little bars on the deck's spindle, so that the spool turns without slipping (similar to the hub and spindle on cassette decks). For very large spools, where the tape needs to be held very firmly, a more complicated system (right) is used. The hub is similar to that on a cine spool, but much larger in diameter; a correspondingly larger-holed spool must be used. The main difference is that the hub can be locked on to the spool, holding it very firmly, and allowing the deck to be operated in a vertical position. This is the *NAB* hub, and is widely used with 27cm spools. Most decks have adaptors to take NAB hubs.

mated later (*mixed down*) by recording them on to another stereo machine. This allows you to experiment at leisure with how the outputs are balanced. Professional studios, especially when recording rock music, may use tape recorders capable of recording 32 or more tracks – mixed down, perhaps, with the help of a computer.

Most decks will take **spool sizes** up to 27cm (10½in). Some will take only 18cm (7in) spools – not so good either for studio work (because the high speeds used means that you run out of tape very quickly) or for most domestic work (where you may want the longest running time you can get). Beware of decks that offer reverse operation: this is usually on replay only, and so is of relatively little use. More details on reel-to-reel tape on page 116.

Tape counters

Counters on reel-to-reel decks are often more accurate than those on cassette decks. They need to be for studio work, where tapes must be quickly cued up for recording and editing. For domestic use, high accuracy is no more necessary than on cassette decks. Most are *linear*: equal counter distances mean the same length on the tape, however much of the reel has passed through – and so, many are marked either with elapsed time (though these will be correct at only one speed) or length of tape. Elapsed time would probably be of most use in studio work, where you may well want to get a rough idea of how long a piece of recording is. For domestic use it is less necessary that the counter numbers mean anything in terms of time or distance, or indeed that the counter is linear.

Tape type switches

It is rare to use anything but ferric tape. There are relatively few brands of this, and the variation in bias between brands is not huge. So there is little need for the complex bias adjusting arrangements that cassette decks must have. However, a few decks have switches for a reel-to-reel ferrichrome tape, and all will have at least internal presets for setting up bias. Some have external presets so that users can set up bias easily themselves.

For studio work, it is essential that the deck is accurately aligned for optimum bias, head alignment and many other factors – not only so that the very best recording will be made, but also to ensure that the recordings will replay on other decks equally well. So, for this sort of use, you would want to make sure that alignment was easy, and that you had the test equipment (and knowledge) to do it properly. For domestic use, such precision is probably not necessary – but it makes sense, if you are going to spend nearly a thousand pounds, to have the deck carefully aligned when you buy it for one brand of tape, and to have alignment checked regularly.

Switches for *equalisation*, may be necessary. There are two reel-to-reel record/replay equalisation standards – *IEC/DIN* and *NAB*. If you want swap tapes with other recorders, especially at professional level, you really need switching for both standards. If you don't intend to swap, then either will do, though the IEC/DIN equalisation gives less hiss on replay. All this applies at 38cm per sec; at 19cm per sec, almost everyone uses NAB; at 9.5, the two standards are the same.

Recording level indicators

Perhaps surprisingly, meters are often little better than those on cassette decks – under-reading on peaks; perhaps over-reading on steady tones. For some studio work, this may not be important: most of the information about levels may come from meters fitted to a microphone mixing desk – but for any other sorts of work, you may need to look for a deck that has got good metering.

Level controls

Most decks have both record and playback controls. For studio work, the most important thing is that the controls are easy to operate – nice for domestic work, too, though less essential because you would not be likely to do so much knob-twiddling. Some decks have *mixing facilities*, which could be useful for the odd live recording session, but for serious studio work, you are likely to want a separate mixing desk which can be tailored to your needs.

Sockets

Most decks have the usual range of DIN or phono input and output sockets; the more professional decks might use different, professional-type connectors. For domestic work, there should be little problem in matching inputs and outputs; for studio work, it may be important to ensure that the input, output and overload levels are all very well chosen. You'll need to check reviews of the decks to find this out.

Editing

If you want to do much editing on a deck, look for features such as a *splicing block* built into the top panel; *easy access* to the heads for marking cutting points; *cue and review*-type operation; the ability to *turn spools by hand*; and generally *good ergonomics*. Ask people who have used the deck you intend to buy for their opinion of ease of editing.

Reel-to-reel tape

Reel tape is ¼in wide – nearly twice as wide as cassette tape – and wound on a single spool. To feed the deck, the free end of the spooled tape has to be put by hand around various guides and rollers, and attached to an empty spool which fills as the tape runs through the machine. The non-magnetic side of the tape may be shiny or matt in finish; matt tapes generally wind more neatly and so last longer.

Because there are different speeds and track formats, different tape thicknesses and spool sizes, there is no simple guide to playing time. Of the several reel sizes available, 27cm or 18cm is probably the best choice. The smaller sizes are less economical and do not give very long playing times. There are also four thicknesses to choose from. The thickest is single-play, and is used mainly by professionals; the thinnest is triple-play and like cassette C120s, the tape is really rather thin and not recommended unless you need the extremely long playing time that it can offer. The other thicknesses are double play (DP) and long play (LP). LP will give better print-through; DP will give longer playing time.

The choice of reel tape is less difficult than for cassette tape. Brands differ in quality and value for money – so study reviews to find a brand you like and have your machine set up for it. Use 27cm or 18cm spools for best economy, and either LP or DP thicknesses.

Although there is likely to be little difference in sound quality between a good cassette deck and a reel-to-reel machine, these are the areas where the differences might be greatest.

Distortion. Reel-to-reel decks should have less distortion at high frequencies – and in general, they should be better at coping with over-recording than are cassette decks. There is less high-frequency compression, too, especially at high speeds.

Reel tape playing times

Spool diameters		Tape length		Playing time (minutes)		
cm	in	m	ft	9.5cm per sec	19cm per sec	38cm per sec
Long play						
8	3	60	210	11	5.5	2.75
10	4	135	450	22.5	11	5.5
13	5	270	900	45	22.5	11
15	5¼	365	1200	60	30	15
18	7	545	1800	90	45	22.5
27	10½	1095	3600	180	90	45
Double play						
8	3	90	300	15	7.5	3.75
10	4	180	600	30	15	7.5
13	5	125	1200	60	30	15
15	5¼	545	1800	90	45	22.5
18	7	730	2400	120	60	30

All times and lengths are approximate. Conversions between imperial and metric measurements are rounded. Single-play tape lasts half the length of double-play tape for the same spool size; triple-play lasts twice the length of long play.

Stability. Reel-to-reel recordings are likely to be free from even the least trace of wow and flutter. There should be less drop outs, too, and head to tape contact should be better. All these sort of things should lead to a more steady-sounding recording.

Hiss. High-frequency hiss is likely to sound cleaner and smoother, though – without Dolby – probably louder than on a cassette deck. Even so, you may consider it less obtrusive. There may be more mid-frequency hiss though.

Chapter 10 Headphones

Headphones are widely ignored by books on hi-fi, and hardly given their fair share of space in hi-fi magazines. Perhaps the main reason for this is that there is not much sensible advice that can be given on how to choose a pair: probably more than any other piece of equipment, you are on your own.

There are a number of reasons for this. Headphone listening is different from loudspeaker listening – and whether some of the differences count as advantages or disadvantages depends to a large extent on your own needs. There are different types of headphone – and again, the type that is best varies from person to person.

One of the most important points in choosing headphones is that you should find them comfortable to wear. Comfort is not only subjective, but also rather personal: what suits you may not suit someone else. Finally, as most headphones suffer from different sorts of sound quality fault, the sound quality is also rather personal.

The pros and cons

Listening to hi-fi on headphones is totally different from listening via loudspeakers. Some of the differences are clear advantages; some just as clearly are disadvantages. The remainder are advantages to some, disadvantages to others.

These are the main **advantages**.
● Headphones can give a better sound quality – at least in some respects – than loudspeakers can. This is mainly because, being smaller than loudspeakers, they place fewer demands on the amplifier. The main advantage is that distortion, even at relatively high levels, is more or less negligible. A single drive unit can cover the whole frequency range – so the sound problems that cross-over units may give with loudspeakers is eliminated.
● Because headphones are worn close to the ear, they can produce high sound levels without large amounts of amplification. (Indeed, an amplifier is often not necessary at all: you can plug phones directly into a tape deck.)
● Headphones are usually much cheaper than loudspeakers, and, of course, take up much less room.
● For amateur recordists, headphones are essential for monitoring. For many recordings, it is not possible to check on loudspeakers what is being recorded – for example, when recording out of doors, or indoors when the recorder has to be in the same room as the action.

These are the main **disadvantages** of headphones.
● They are often uncomfortable to wear.
● The fact that you have to be within the headphone-lead's length of the amplifier can be rather inconvenient. Some types also need a mains supply. (Of course, you can always extend leads if necessary – and occasionally, a manufacturer brings out a wireless type of phone.)

There are three **other considerations** – advantages or disadvantages, depending on you.
● The stereo effect you get when listening on headphones is completely different from that when listening via loudspeakers. For a start, the sound sources approach your ear from the sides, rather than from much nearer the front. The sound is also much more direct, with no reflections from the listening room (one of the points that can make headphones better than loudspeakers, especially for monitoring.) So most people find that the stereo effect is reduced to a straight line, with the instruments appearing inside the head – which can appear very unnatural. On the other hand, the stereo separation from headphones is usually very good, clear and detailed. Some people find this enhancement more than compensates for the other disadvantages – and it can certainly be a great help when monitoring.
● Listening is very much a private experience. Only one person can listen on each pair of headphones. External

sounds cannot get in, and the sounds from the headphones cannot get out (more or less, depending on the type of headphone). If you want to listen to your music loud, without annoying the rest of your family or your neighbours, this privacy is very useful. It can also mean that you are able to listen to your hi-fi in the same room as any other household activity – without you disturbing them or them disturbing you. But of course, if you want to share your musical experiences, you will find headphone listening very lonely.

● Headphones can easily be driven to very loud volumes. This is useful if you are on a tight budget; you need buy only a small amount of amplifier power. And, particularly for rock music, you might like the sensation of very loud music. But very loud sounds can easily damage the hearing, and headphones can fool you into thinking that you are listening at a much lower volume than in fact you are doing. So it is wise to exercise considerable care in listening, and restraint in setting the volume knob.

How headphones work

Many of the simpler types of headphone are moving coil – roughly, miniature versions of loudspeaker drive systems. But because headphones do not suffer from the restrictions of sheer size that limit loudspeakers, their designers can use a much greater variety of drive systems – so there are all sorts of electrostatic and electromagnetic headphones available.

As usual, no one method is necessarily better than any other: all types of drive can produce good sound if the headphone has been designed properly.

Types of seating

To be heard properly, headphones have to sit against the ears. The two earpieces are attached by a curved springy band which fits over the top of the head, so that the earpieces are forced on to the ears.

There are two main ways of seating the earpieces. They can simply be pressed up against the ears (usually via foam pads of some sort) – the **supra-aural** type. Or

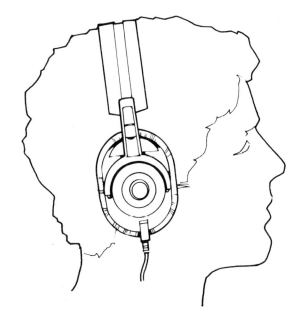

Circum-aural headphones

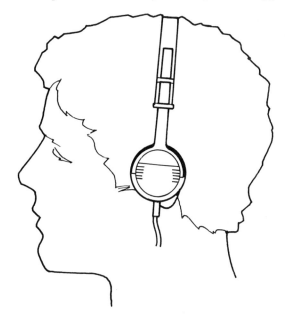

Supra-aural headphones

they can be surrounded by a pad which presses against the head round the ear – the **circum-aural** type.

The type often affects how much **isolation** the phones give – that is, to what extent external sounds can be heard by the headphone wearer, and how much other people can hear what is being reproduced over the headphones. Circum-aural types *tend* to give the most isolation – though there are specific examples that give very little. Supra-aural headphones rarely give much isolation. The majority of designs available at the moment are supra-aural that give little sound isolation.

Whether you want a pair of headphones that has much or little isolation depends on what you want to use them for. High isolation types are used when you want to exclude the outside world as much as possible – listening to the hi-fi in the domestic living room, for example, or for concentrated monitoring of live recordings. On the other hand, headphones with little isolation allow you to keep half an ear open for the doorbell or whatever; and you may also find the sound quality more natural, with less of a shut in feeling.

Binaural sound

Humans are quite good at deciding which direction a particular sound is coming from. The brain uses a large number of audible clues when making up its mind about sound location – the difference in loudness between the sound arriving at the right and left ears, the difference in time taken for a sound to arrive at the two ears, and so on. This happens not only for the sound coming directly from the instrument or whatever, but also for the myriad of reflections and reverberations coming from the wall, floor and ceiling of the room where the instrument is being played. Unless we have our eyes shut, the brain makes use of visual clues, too.

Conventional two-channel stereo reproduced over loudspeakers cannot handle more than a very few of these clues. But luckily, those that it can cope with are usually sufficient to build up a reasonably good illusion of a solid sound, with all the instruments located in their apparently correct positions.

One way of making the illusion more real is to try to capture the exact sounds that a particular listener in an audience (say) would hear individually in each ear – by fitting the listener's ears with small microphones. If a recording of the output of each microphone (which can be made with a conventional stereo recorder) is later replayed *at each ear* (through headphones) of a second listener, they should be able to hear exactly what the original listener was hearing. This is **binaural sound**.

Using a live human for recording purposes is awkward, so a dummy head is usually used; binaural sound is often called **dummy head stereo**. There is much argument about how life-like this dummy head needs to be: some say the closer in shape, density and texture the head and ears are to a human's, the better the results; others say that a head is not necessary at all, and that good binaural recordings can be made simply by using a pair of microphones spaced apart by about the diameter of a human head, with a thick plastic baffle board in between them.

The reproduced effect is not quite as good as the real live sound is – and for some people the effect doesn't work at all. At its best, though, binaural sound is far more effective than conventional stereo ever is, so it is worth trying. The BBC broadcasts special binaural plays from time to time. It would be worth making a high-quality recording of one. Then you can experiment with playing it back through different types of headphones (which might make a difference to the stereo effect) and even through loudspeakers. (Try them positioned either side of you, close to your head.) Although different types of headphone may give slightly different results with binaural recordings, it is *not* necessary to have a special type; any headphone will do.

Comfort

Headphones are potentially uncomfortable – primarily because of the springiness that keeps the earpieces in place against the side of the head. Neither supra- nor circum-aural types are inherently more comfortable: size, weight, tightness and so on are more important. The only real way to find out how comfortable a pair of phones is, is by wearing them – preferably for several hours. You're obviously not likely to get an opportunity to do this in a shop, particularly as headphones cost relatively tiny sums. So where else can you turn to for advice?

Friends may be a good source of several different pairs of phones, which you would be able to try out over a long period. Individual reviews in magazines are likely to be little help, though comparative reviews on comfort are rather better.

Of course, if you intend to use your headphones only infrequently, and for short periods of time, comfort may not be particularly important to you.

Sound quality

About the only technical test worth carrying out on headphones is **frequency response**. But it is not easy to measure the frequency response from an earpiece as it would be heard by the ear. For one thing, the actual response from the earpiece will be modified by the shape of the ear. The usual way round this is to mount the phones on some sort of artificial ear or dummy head.

The second problem is that it is not certain that the output from a headphone when heard by an artificial ear should be flat in order for a real ear to perceive it as flat. Some of the research suggests that, particularly at higher frequencies, the best response is anything but flat.

All this makes it difficult to interpret published frequency response charts – whether in reviews or in manufacturers' specifications.

Headphones do seem to vary a lot in response, which in one sense makes it even more difficult to interpret response charts. How do you rate the relative merits of say, a 5dB dip at 3kHz together with a roll-off above 8kHz against a bass roll-off below 150Hz together with a 6dB peak at 10kHz?

The answer, of course, is a **listening test**, carried out as part of a *comparative* review.

Headphones can be a useful alternative to loudspeakers. In many cases, cheap ones seem to perform as well as some of the more expensive ones – so there is no need to spend more than about £15.

Look for a pair that is comfortable to wear (especially if you intend to wear them for long periods), and provides the amount of isolation that you want.

Frequency response varies a lot from model to model; listening tests (either yours, or those of a good comparative review) are the best way to pick a pair that give the flattest apparent response. Published frequency response curves may be misleading.

Be careful not to listen too much at even moderately loud levels.

Chapter 11 Microphones

Most owners of tape recorders and decks are content to record (with or without permission from the copyright holders) from discs or from the radio. Those who want to make their own live recordings will need to use microphones. There used to be a time when all tape recorders came with a microphone – even if not a very good one – but nowadays, most cassette decks do not – though all will accept mics.

There are various types of microphone, with different characteristics, used for different types of recording. This chapter is a basic guide to the different types. If you want to go further – into serious music recording, perhaps, or into specialisms like how best to capture the sound of bisons mating, there are books on the subject. Or, probably best, join a tape recording club. Members can pool their equipment and experience.

FEATURES ○|○|○|○|○|

Polar response
The electrical output from some types of microphone varies depending on what direction the sound is coming from. A way of showing this in pictures is to use *polar diagrams* – the sort of circular graph shown opposite. What you are trying to record determines the type of polar response you go for. For example, for recording a boardroom conference, an **omnidirectional mic** would ensure that all the directors round the table would be heard equally well. A **bidirectional mic** is useful for face-to-face interviews. A **cardioid mic's** response is more directional, useful for cutting out unwanted sounds – for example, trying to record a pub band without too much pub getting into the act.

Looking at it another way, the more directional the microphone, the further away from the subject it can be without the sound getting lost in unwanted background noises. Slightly more directional than the cardioid is the similar **hyper-cardioid**. More directional still is the **rifle mic**. (Where it is necessary to pick up very quiet sounds from relatively far away, a microphone can be fitted with a parabolic shield – but these give a poor bass response, so their use is limited to things like bird song and spying.)

Impedance
The output impedance of a microphone depends mainly on the type – though the impedance can often be changed with matching transformers, which are sometimes built into the microphone itself. There are three main levels of impedance: *low* (about 30 to 250ohm), *medium* (250 to 1,000ohm), and *high* (around 25kohm). Microphones may be marked with their impedance, or just low, medium or high. Tape recorder specifications will usually give the input impedance of their mic input: in general, try to use a mic with an impedance three to five times *lower* than that input impedance. Matching a low impedance microphone to a high impedance input should be all right – but the other way round is likely to give distortion.

If you want to run long leads from a microphone to your recorder, you will have to use low impedance types; high impedance ones used with long leads results in loss of treble and increase the chance of interference.

Sensitivity
There are many different ways of expressing the sensitivity of a microphone – that is, how large the electrical output is for a given loudness of sound. In general, there is little need to worry about it, but try to arrange a trial before buying.

MICROPHONES

These polar diagrams are simply circular graphs, and can be read like any other graph. The starting point is the centre of the circle; points on the graph to the left and right represent what happens to the left and right of whatever is in the centre; points above represent the front, and below represent the back. The various circles represent amounts: the greater the diameter of the circle, the larger the amount it represents. If the polar diagram is to represent a microphone's output, the amounts will be different output levels, in volts.

Other polar responses can be made by adding baffles and shield to a microphone, or by adding the response of one type to that of another.

Microphones – poorer ones especially – can have slightly different polar responses at different frequencies. This can lead to instruments appearing to fade or grow in volume, or change position, as they play different notes.

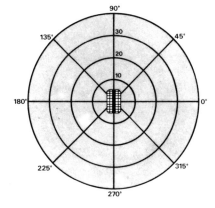

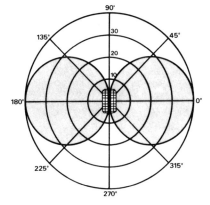

An **omnidirectional microphone** gives the same level of output whatever direction the sound approaches from. The curve on the graph is at the same distance from the centre for all angles (ie it is a circle).

A **bidirectional**, or **figure-of-eight microphone**, whose output is zero for sounds that approach from either side. As the sounds swing round to the front or back, though, the output from the mic increases; it is at a maximum for sounds directly from the front or back. The polar response curve has the shape of an eight.

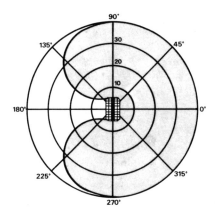

A **cardioid microphone** (left) has an output that is at a maximum only for sounds from the front. As the sound swings away to either side, the output from the mic decreases – slowly at first, but rapidly to zero as sounds reach the back. The polar response is heart-shaped.

A **hyper-cardioid microphone** is a variation which gives some output from the very back, but less from the back sides.

Microphone techniques

Knowing the difference between crossed pair and multi-miking won't turn you immediately into a recording engineer. But it will help you understand something of what record reviewers are talking about in the technical sections of their reviews.

Crossed pair (also called *Blumlein* or *co-incident*). Two directional microphones are used, placed as close together as possible. One points towards the left-hand side of whatever is being recorded, the other to the right-hand side. The angle between the directions that the two mics are pointing in is usually 90° or so. Sounds directly from the left side are picked up strongly by the left-hand mic; those from the right by the right-hand mic. Sounds from the centre are picked up by both microphones equally.

Many recording engineers find this technique extremely good, especially for basic classical orchestral recording. The stereo effect is particularly pure and clear. A great amount of skill is needed to know exactly where to place the crossed pair to get the best result. Altering the recording balance, if you have got it wrong, is difficult, and usually means re-positioning the microphone, which is often suspended in mid-air. (Special mics are available whose apparent position can be changed electronically at the control desk.)

Even the purist recordist will need to use additional microphones to record some works – for example, a choir and separate organ, or large choir and orchestra.

Coincident pair recording is not really suitable for non-classical work – especially rock or pop music.

Spaced microphones. This technique uses two basic omnidirectional microphones placed many feet apart – one on the left and the other on the right of the area. The technique is widely used in classical recording, particularly in the USA – but many critics find that it produces an unclear sound, with poor positioning, especially from the centre of the orchestra. Inevitably, parts of the area are not well covered by the basic two microphones, and others – often many others – have to be added.

Multi-miking. This is virtually the only technique used for non-classical recording (and for some classical, too). Each instrument, or group of instruments, is given its own microphone – mono or, more rarely, stereo. The sound from electronic instruments – organs or guitars – may be taken directly from their electrical outputs, or via a microphone, perhaps up against their loudspeakers.

All the mics and other outputs are fed to a control desk. Here, they can be altered in volume from very loud to non-existent; they can be positioned anywhere across the stereo stage from extreme left to extreme right; their frequency response can be changed electronically; special effects can be added. Groups of outputs are added together and fed to a multi-track tape recorder. The more tracks on the tape recorder, the fewer decisions the recording engineer and record producer have to make during the recording session. Exactly how to balance, position and shape the overall sound can be sorted out later when the tapes are replayed and re-recorded.

The result is obviously very artificial, but much rock music has no real existence to compare this with. For classical recording, there tends to be more reality – and many critics find that multi-miked classical recordings have a generally poor technical quality. This is mainly because each microphone picks up not only the instrument it is intended to pick up, but also instruments intended for other microphones – quieter, and with some time delay – which can lead to an ill-defined sound, with poor stereo positioning and transients.

Even in classical recording, though, multi-miking gives the engineer more flexibility to change the balance of each instrument in volume and positioning without having to reposition microphones.

Types of microphone

There are three types of microphone you are likely to choose from.

Moving coil (also called dynamic). They give low distortion and a reasonable frequency response, and can be cheap. They are also robust. Their main fault is that they can give a rather coloured sound with a peak in the mid to high frequencies – though this can be an advantage, for example when recording a voice out of doors.

You could use a moving coil mic for recording most things, but its colouration may be obtrusive on woodwind and string music.

Moving coil microphones have an omnidirectional or cardioid response (the cardioid is probably more useful); any impedance (but usually medium); reasonable sensitivity.

Ribbon. These have been widely used in broadcasting. They can give an excellent sound quality, though perhaps rather mellow, without the resonance of the moving coil type. Their main fault is that they are very fragile. The best ribbon mics are rather expensive.

You could use one for speech (especially male voices, which they tend to flatter), or for music – especially classical woodwind and string sections. They cannot be used out of doors: even light breezes can damage them.

They normally have a bidirectional response, though with some models this can be modified; a low impedance (even then they need a built-in transformer); low sensitivity.

Capacitor (also called a condenser). This is by far the most common type used by professionals, mainly because the sound quality is very good especially at high frequencies. It is reasonably robust, though not too keen on damp conditions. It is very expensive though, and requires an external DC power supply – though some models can operate off batteries.

The similar *electret* type gets round some of these problems. They are cheaper and do not need an external power supply. (They still need an internal battery, but the drain on this is so low that its life is very long.) But the sound quality is often not as good, with many peaks in the response. Some electrets have a limited, though long, life; after a while, their sensitivity gradually decreases.

Capacitor microphones can be used for most types of recording, including out of doors, and are particularly useful for recording music with lots of high frequencies – brass, for example.

Most models have an omnidirectional or cardioid response – though some of the more expensive types have a response pattern that can be varied, perhaps even remotely; output impedance is medium; they usually have a high sensitivity.

For live recording, you will need a microphone or two. If all you want to do is to record family events such as baby's first words, or a speaking letter to send to the folks back home, then any cheap dynamic-type microphone will do.

If your interests lie in more serious recording, then your choice of microphones requires more thought, and preferably a working knowledge of what each type does. If you can, join a club and learn from those who already have the knowledge.

Index

A

aerial input sockets	84, 91
aerial rotator	84
aerials	84, 91–94
amplifier	
built into loudspeakers	37
description	9
integrated vs separate	47
transistor	36
use with headphones	118, 119
valve	36
amplitude, sound waves	18, 20
amplitude modulation	80, 81
ANRS noise reduction	104
anechoic chamber	38, 43
Adres noise reduction	104
arms – *see* pick-up arms	

B

balun transformer	84, 91
BBC	81, 94
bias	
recording	96, 102
tape	97, 109
bias compensation	74
binaural sound	120
broadcasts, stereo	80

C

cables	
loudspeaker	44
microphone	122
capture ratio	88
caring for discs	78
cartridge, pick-up – *see* pick-up cartridge	
cartridge tape recorders	96
cassette deck	
azimuth	107
cleaning	110
controls on amplifier	51, 52
description	9
re-alignment	97
matching to tape	97, 107
v. reel to reel tape deck	112, 115
cassette tape	
drop outs	109, 110
equalisation	98
matching to deck	97, 107
playing time	98
print through	110
speeds	98, 101
types	98
vs discs	100
colouration	40
complex load	58
compliance	65, 66, 70, 77
compression, high frequency	106
copyright, recording	96, 122
crackles, tracing	16
cross talk	
amplifiers	58
cassette decks	108
pick-up cartridge	77
tuners	90
(*see also* stereo separation)	
current, electric	25

D

dbx noise reduction	104
dealers	15, 59
decibels	20
demonstrations	15
digital recording	22
DIN	
plugs and sockets	52
Standards	13
weighting curve	31
discs	
cleaning	78
vs cassette tape	100
distortion	
and signal to noise ratio	29
cassette deck and tape	103, 108
clipping	30
clipping, amplifier	53, 54
crosstalk	90
disc	17
harmonic	32
harmonic, amplifiers	55
intermodulation	32
intermodulation, amplifiers	55
intermodulation, tuners	90
loudspeakers	43
measurement	32

INDEX

multipath   83, 89, 92
pick-up cartridge   77
RFIM   89
reel to reel decks   116
tracing in record decks   64
transient harmonic   55
transient intermodulation   55
Dolby
    alignment   97
    level   31, 107, 109
Dolby noise reduction   96, 107, 109
    Dolby A   96
    Dolby B   96, 104
    Dolby C   105
    Dolby HX   105
drop outs, cassette tape   109, 110
dynamic range   29
    cassette tape   108, 109
    orchestra   21
(*see also* signal to noise ratio)

E

earth loops   17
earthing   17
editing
    reel to reel   116
effective arm mass   66, 71
eigentone   43
electricity, basics   25
equal temperament   21
equalisation, tape   96, 102

F

filters
    amplifier   56, 57
    pilot tone and sub carrier   90
flux levels   107

frequency
    intermediate   89
    sound waves   18
frequency compression   106, 108
frequency modulation   80
frequency response   27
    amplifier   28, 55, 56, 57
    amplifier, cartridge matching   59
    and signal to noise ratio   31
    cassette deck   106, 107
    cassette tape   109, 110
    headphones   121
    loudspeakers   29
    pick-up cartridges   28, 59, 76
    record decks   69
    test methods   28
    tuner   90
fuses   16

G

graphs, how to read   26

H

headphones vs loudspeakers   118, 121
heads, tape decks   95, 103, 107, 108
headshell, pick-up arms   66
hi-fi
    cost   15, 16
    definition of   9
    magazines   13
Hi-fi Choice   13
    turntables and tone arms   72
    cartridges and turntables   71
    loudspeakers   39
Hi-fi For Pleasure, cassette
    decks and tapes   97

high frequency compression   106, 108
hiss, reel to reel tape decks   116
(*see also* signal to noise ratio)
hum
    record decks, reducing   17
    tracing   16
    tuners   90

I

IBA   94
IHF, amplifiers   57
impedance
    electrical   25
    loudspeaker   40, 53
    microphone   122
    pick-up cartridge   69
    sockets   52
inputs, types   52
interference, tuner   88, 92
intermediate frequency   89

L

leads – see cables
limiting, tuners   86
listening position   18
listening rooms   41
    acoustics   43
listening tests
    amplifiers   46
    cassette decks   109
    cassette tapes   97
    headphones   121
    hints on   14
    loudspeakers   40
    record decks   61, 70
loading, cartridges   69
loudness   20, 22

INDEX

loudspeakers 39
loudspeakers 10
   as amplifier loads 58
   controls on amplifiers 52
   music centres 11, 12
   positioning 41
   vs headphones 118
   with built-in amplifiers 37

M

matching
   amplifier and cartridge 59, 69
   cartridge and pick-up arm 70
   cassette deck and tape 96, 107
   loudspeaker and amplifier 38, 40
   loudspeaker and room 38, 39, 41
   record deck and room 70
maximum output level, cassette tape 109
microphones, controls on amplifiers 52
MOL, cassette tape 109
monitor loudspeakers 37
monitoring recordings 103
mono 9
MPX filters
   cassette deck 105
   tuner 90
music 21
music centres 11, 12
music power 54

N

noise – *see* signal to noise ratio
noise reduction on cassette
   decks 104
   Adres 104

ANRS 104
dbx 104
Dolby 96, 104, 105, 107, 109

O

offset angle 72
output level
   pick-up cartridge 62, 63
   tuner 84
outputs, types 52
overhang 72

P

phase 20, 26
   loudspeakers 35
phase compensation, loudspeakers 37
phono plugs and sockets 52
pick-up arms
   effective mass 66, 71
   headshell 66
   matching to cartridge 70
   setting up 71
   types 65
pick-up cartridge 59, 63
   compliance 65, 66, 70, 77
   matching to amplifier 69
   matching to arm 70
   playing weight 73
   tests 76
   trackability 77
pilot tone 80, 90
pitch 18, 21
plugs
   loudspeaker 44
   types 52
polarizing voltage

loudspeaker 35
   microphone 124
power
   electrical 25
   music 54
power bandwidth 54
power output, amplifier 46, 53, 119
power rating, loudspeakers 38
price of hi-fi 15
print through, cassette tape 110

Q

quadrophonic
   controls on amplifiers 50
   Hafler type controls 50
   wiring loudspeakers 45
quadrophonics 9

R

rack systems 11
record deck
   controls on amplifier 51
   positioning 70, 75
record decks 9
recording level
   discs 62
   tuner 84
records – see discs
reel to reel tape decks 9
   cleaning 110
   v. cassette decks 112, 114
repairs, diy 16
resistance 25
resonances
   loudspeaker 40, 42, 43
   pick-up cartridge 71
   record deck 61

INDEX

RIAA
   correction   62
   recording characteristic   55
rms   26
rooms, listening   41, 43
rumble, record decks   75
rumble weighting, record decks   75

S

scale, musical   21
selectivity   83, 87, 92
sensitivity   33
   amplifier input   49
   cassette tape   97, 109
   loudspeaker   38
   microphone   122
   tuner   85, 86
separation – see stereo separation
signal to noise ratio   29
   amplifier   57
   cassette decks and tape   31, 103, 108
   reel to reel tape decks   116
   test methods   29
   tuner   83, 86, 89
   (*see also* dynamic range)
sockets
   amplifiers   51
   cassette decks   106
   loudspeakers   44
   types   52
sound   18
sound level
   and headphones   119
   maximum   39
sound waves   19
   amplitude   6
   frequency   18
speakers – see loudspeakers
spectrum analysis   32

speed accuracy, record decks   68, 75
speeds
   cassette deck   101, 109
   record deck   68, 75
   reel to reel tape deck   114, 116
spools, reel to reel   114
Standards   13
standing waves   43
stereo   9
   dummy head   120
   with headphones   118
stereo image, checking   44
(*see also* cross talk and stereo separation)
stereo separation
   amplifiers   58
   cassette decks   108
   pick-up cartridges   77
   tuners   90
stereo switching point   86
stroboscope   68
stylus
   care and cleaning   78
   life   65
   pick-up   63
   setting up   71
sub carrier   80, 90

T

tape – *see* cassette or reel to reel
tape care   111
tape deck – *see* cassette deck or reel to reel tape deck
tilt, stylus   73
time-delay compensation   37
tone colour   21
tower systems   11
tracking error, pick-up arms   64, 72
tracks
   cassette tape   98

reel to reel tape   114
transient response, amplifiers   55
transients   23
tuner   9
tuner/amplifier   12

U

upgrading   16
   racks   12

V

vertical tracking angle   73
vibrations in record deck systems   61, 70, 76
voltage, electric   25

W

watts   25, 53
wavebands, radio   80
waveforms   23
where to buy, advice   15
Which?
   amplifiers   46, 55
   cassette decks and tapes   97, 107
   demonstrations   15
   hi-fi   13
   loudspeakers   37, 38, 39, 43
   music centres   11
   pick-up cartridges   69, 71, 77
   record decks   59, 61, 67
   tuners   80, 88
wiring, loudspeakers   44
wow and flutter   33
   cassette deck   108
   cassette tape   110
   record decks   75